An OPUS book

POLITICS IN THE MIDDLE EAST

Politics
in the
Middle East

ELIE KEDOURIE

Oxford   New York
OXFORD UNIVERSITY PRESS
1992

Oxford University Press, Walton Street, Oxford OX2 6DP

Oxford New York Toronto
Delhi Bombay Calcutta Madras Karachi
Petaling Jaya Singapore Hong Kong Tokyo
Nairobi Dar es Salaam Cape Town
Melbourne Auckland

and associated companies in
Berlin Ibadan

Oxford is a trade mark of Oxford University Press

British Library Cataloguing in Publication Data
Data available
ISBN 0–19–219167–5
ISBN 0–19–289154–5 Pbk

Library of Congress Cataloging in Publication Data
Kedourie, Elie.
Politics in the middle East / Elie Kedourie.
p. cm.
'An Opus Book.'
Includes bibliographical references and index.
1. Middle East—Politics and government. 2. Turkey—Politics and
government—1909– 3. Middle East—Constitutional history.
I. Title.
956—dc20   JQ1758.A2K4   1992   91–27206
ISBN 0–19–289154–5
ISBN 0–19–219167–5

Typeset by Cambridge Composing (UK) Ltd
Printed in Great Britain by
Biddles Ltd.
Guildford and King's Lynn

Acknowledgements

I WROTE this book in various intervals, when free from teaching and other duties. I am most grateful to the Netherlands Institute for Advanced Study; to the Sackler Institute of Advanced Studies, Tel Aviv University; and to All Souls College, Oxford, for having, with very great generosity, afforded me the leisure to do so. I began the book while a Fellow of the Netherlands Institute in 1980–1 and put finishing touches to it while a Visiting Research Fellow at All Souls in 1989–90.

These dates will indicate why I am also particularly grateful to the Oxford University Press for the great patience they have exercised—patience beyond the call of any obligation—while waiting for me to complete the work.

E. K.

For GEORGE

By the same author

England and the Middle East: The Destruction of the Ottoman Empire 1914–1921

Nationalism

Afghani and 'Abduh: An Essay on Religious Unbelief and Political Activism in Modern Islam

The Chatham House Version and other Middle Eastern Studies

Nationalism in Asia and Africa

Arabic Political Memoirs and other Studies

In the Anglo–Arab Labyrinth: the McMahon–Husayn Correspondence and its Interpretations 1914–1939

Islam in the Modern World and other Studies

The Crossman Confessions

Diamonds into Glass: The Government and the Universities

Perestroika in the Universities

Edited by the same author

The Middle Eastern Economy

The Jewish World

Edited by the same author with Sylvia G. Haim

Modern Egypt

Towards Modern Iran

Palestine and Israel in the 19th and 20th Centuries

Zionism and Arabism in Palestine and Israel

Essays on the Economic History of the Middle East

Contents

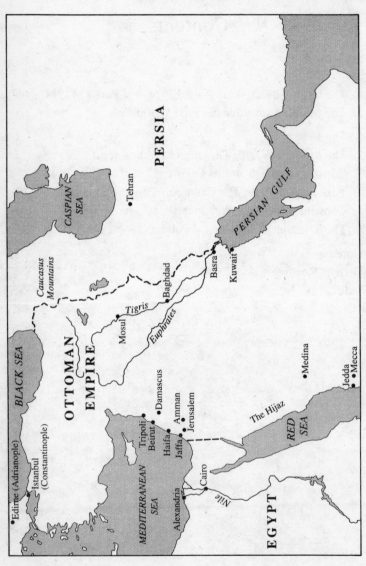

The Ottoman Empire in Asia, Egypt and Persia in 1914

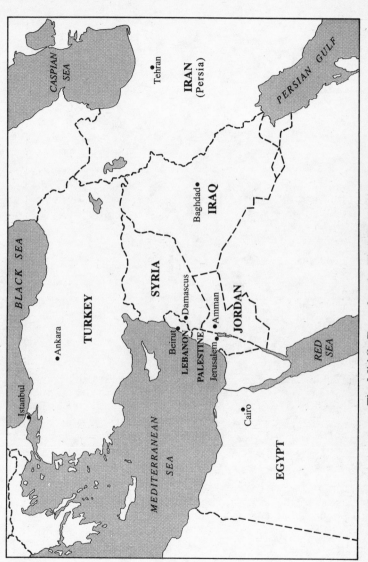

The Middle East after the First World War

1

The Legacy

WHETHER it is defined in geographical or cultural terms, and whatever its exact boundaries are held to be, there can be no disputing the fact that the Middle East is predominantly Muslim. The beliefs, norms, and attitudes of Islam, the experiences, triumphs, and vicissitudes that Muslims have encountered over the centuries have combined to bring about a society of a highly distinctive character, with its own unmistakable patina. Muslims, but equally non-Muslims dwelling in the Muslim domain, are strongly marked by the Muslim tradition, and what may be called the Muslim civilization, the more so that until very recent times, the Muslim world had little contact with, little curiosity about, and little respect for what went on outside its boundaries. Even today, when the Western world is the source of industrial techniques and military weapons, and is seen as providing intellectual and political norms, Islam as a religion is very far from being defeated or silenced. And its influence as a culture, whether acknowledged or not, obstinately persists in permeating and shaping institutions, attitudes, and modes of discourse. This is nowhere more true than in government and politics, and in the mutual responses of the rulers and the ruled.

In the version of Islamic history traditionally accepted by Muslims, Islam, as the Prophet Muhammed left it at his death in the year 632 of the Christian era (CE), was not only a religion, but also a government, and a fast-rising international power. Tribes and tribal chieftains were converted to the new religion and made their submission to the Prophet, who was at once the Messenger of God, and the political head of the new Muslim community—a political head who was also leader in war and justiciar. The vast territories conquered, from very soon after the Prophet's death, were conquered and administered in the name of Islam, and became a part of the Islamic domain. What Western Christianity from very early times distinguished as temporal and spiritual were *ab initio* inextricably intermixed in Islam. Islam made the Muslims one community where all concerns, spiritual and temporal, were attended to and codified in the Divine Law as revealed in the Koran and in the Prophetic Traditions. This community, the jurists taught, constituted the Muslim *umma*, i.e. a body of people who were the object of the divine plan of salvation and who therefore were the proper unit of political organization. The *umma* is unlike a Greek polis or a modern nation-state. Its basis is not kinship or the occupation of a well-defined national territory. Wherever the *umma* is, there is *dar al-islam*, the abode of Islam; all the rest is *dar al-ḥarb*, the abode of war, where infidels rule. *Dar al-islam*, therefore, knows no permanent territorial frontier, and whatever comes under Islamic authority becomes part of *dar al-islam*. Nor is the *umma* like the Roman empire, a city-state which eventually developed into an extensive dominion, where citizenship became a legal, but not a religious bond. The nearest analogy to the *umma*, but a very imperfect one, is the idea of a *respublica christiana*. But this *respublica* never achieved even the most transient political reality, and from early on it was supposed to be governed by two distinct, albeit related, authorities, the *regnum* and the *sacerdotium*.

The *umma*, then, is a community dedicated to the service of God according to His commandments, and to spreading the true faith. The religion, therefore, is necessarily inseparable from politics. This may be seen from the fact that a perpetual duty which Islam enjoins on the community as a whole is the duty of jihad, that is, of taking arms against the enemy at the behest of the rulers. On the border, in battle areas, and whenever the ruler so orders, the collective duty, however, becomes incumbent on each individual Muslim. The Prophet, as has been said, was considered to be the first ruler, justiciar, and leader in prayer of the religious polity which he had established at Medina. He took the political decisions, received emissaries from the tribes, appointed tax-collectors, and decided on military expeditions. At his death, these functions are said to have devolved on one of his earliest followers, his father-in-law Abu Bakr, who, in the traditional account, came to be known as the successor, the *khalifah*, of the Messenger of God. Whether this was the original significance of the title, or whether, as has been cogently argued,[1] the *khalifah* was originally understood to be God's deputy rather than the Prophet's successor (the word *khalifah* signifying both successor and deputy), those who assumed the office of caliph united all these powers in their own person. Caliphs also had other titles which emphasize various aspects of the office. The caliph is *amir al-mu'minin*, commander of the faithful. This title emphasizes the Muslim ruler's authority as military leader. The immense and astonishing conquests made by the Muslims soon after the appearance of Islam gave this title a particular weight and resonance. Another title given to the Muslim ruler is that of imam, which emphasizes the religious character of his office. The imam in Muslim worship leads the worshippers at the daily prayers, and is followed by them in performing the prescribed ritual movements of bowing, kneeling, and prostration. During his lifetime the Prophet acted as imam for the Muslim community, but in his last illness he is said to have ordered Abu Bakr to act as imam. This is believed to have facilitated Abu Bakr's choice as caliph, since in Muhammad's lifetime leadership in prayer undoubtedly signified leadership of the community. While the prophetic office was naturally peculiar to Muhammad, the maintenance of the Divine Law was a responsibility which devolved upon his successors, along with all the other functions which he is held to have exercised at Medina.

The caliphate or imamate, then, is, as laid down by the jurists, the highest public office in Islam, indeed the only one in which authority, subject of course to Divine Law, is original. The office combines executive and judicial authority in the widest possible sense. At any one time it inheres in one person alone, and the Muslim world did not find it possible to set up any institution or device whereby the members of the *umma* could articulate their varying opinions and interests, and whereby the power of the caliph could be checked, counterbalanced, or in any way contained. This may not have seemed necessary at the beginning, for the caliph could be considered as bound by the ordinances of Divine Law, charged with maintaining it, and with attending to the common concerns of the Muslims. In any case, the office was, at the start, elective, and the election was held to be conditional. In accepting the *bay'a*, the formal act of homage signifying the recognition of his authority, the caliph, as the jurists taught, bound himself to exercise his office within the limits set by Divine Law, and to defend, secure, and generally tend the interests of Islam. This was a legally binding contract of government whereby obedience was due so long as the caliph did not order the Muslims to do what is contrary to God's Law.

The duty to obey the caliph was not simply the outcome or concomitant of a civil contract between ruler and ruled: it was grounded in religion. The Koran declares; 'O believers! Obey God and obey the Prophet and those who are in authority over you.' And many *hadiths*, i.e. traditions attributed to the Prophet and having the same binding force as Koranic revelation, also enjoin this duty. Thus, one *hadith* declares: 'The Apostle of God said: "Whoso obeys me, obeys God, and whoso rebels against me, rebels against God. Whoso obeys the ruler, obeys me, and whoso rebels against the ruler, rebels against me."' The reason for such injunctions is clear: a godly life is a life lived according to the *shari'a*, the Divine Law. But the law is not maintained by itself, it requires someone to uphold and enforce it. As one jurist was to put it:

At the head of the Muslims there must necessarily stand someone who sees to it that their laws are carried out, their statutes maintained, their borders defended, and their armies equipped, who makes sure that their obligatory taxes are collected, that men of violence, thieves and highwaymen are suppressed, that services are held on Fridays and feast-days, that minors (in need of a guardian) can be married, that the spoils of war are justly divided, and that similar legal obligations, which no single member of the community can take care of, are performed.[2]

The vicissitudes of Muslim history, the divisions and schisms which marked its earliest history, the heresies, the civil wars, the insubordinate soldiery, and the disorders they continually generated made the necessity of a ruler who would protect the community and keep it together an ever more insistent and prominent leitmotiv in Muslim political thought. Some five centuries after the appearance of Islam, the great theologian Ghazali (who died in 1111 CE) eloquently set out in a classic passage the reasons why an imam has to be instituted, and why obedience to him is not only a rational, but also a religious obligation:

We maintain that the right ordering of religion was certainly the purpose of the Agent of revelation. This is an absolute premise on which there can be no disagreement. We will add to it another premise, to the effect that the right ordering of religion can only obtain where there is an imam to whom obedience is given. It follows from these two premises, that the contention that an imam must be set up is correct. If it be said however that the second premise is not granted, to wit, that the right ordering of a religion can only obtain through an imam to whom obedience is given, therefore prove it, then we will say: The proof is that the right ordering of religion can only obtain where there is right ordering of the world, and the right ordering of the world can only be obtained through an imam to whom obedience is given. Here are two premises. Where does the disagreement lie? And if it be asked: Why do you say that the right ordering of religion can only result from the right ordering of the world, when, in fact, it can only result from the ruin of the world, for are not the spiritual and the temporal opposites, and does not the prosperity of the one entail the ruin of the other? Our answer is: This is the speech of him who does not understand what we mean by 'the world' in this context. It is an ambiguous word which may, on the one hand, refer to excessive self-enjoyment and pleasure beyond need and necessity, and, on the other, it may refer to all the necessities in the life before death. In one sense, it is indeed against religion, but in the other, it is the very condition of religion . . . We say therefore that the right ordering of religion comes through knowledge and worship, and these can only be attained with a healthy body, with the preservation of life and with the satisfaction of the necessities of cold, hunger, exposure, and insecurity, the ultimate evil. On my life, for him who is secure in his freedom, healthy in body, able to obtain his daily bread, it is as if the world has been fulfilled in all its aspects! . . . Religion cannot be rightly ordered except through the fulfilment of these essential needs; otherwise, how can a man who spends all his time defending himself against the sword of the oppressors, and seeking his food in the face of those who overreach him, how can such a man find time to devote himself to learning and [religious] endeavour, for these are the means to a blessed afterlife? Therefore, the right ordering of the world, that is, according to its necessity, is a condition for the right ordering of religion.[3]

The reader will sense in Ghazali's argument a clear-eyed, sceptical pessimism, or even desperation, about what the ruled can at best expect from their rulers. But in the passage just quoted there is no explicit indication of the fundamental change in the nature of government which had overtaken Muslim society long before Ghazali's time. As has been seen, the relation between ruler and ruled was supposed to be governed by a contract whereby the subject promised to obey the elected ruler so long as the *shari'a* was maintained. Very shortly after the Prophet's death the Muslim domain expanded enormously. Within a few decades it came to include Egypt, the Levant, Mesopotamia, Persia, and Central Asia—areas of high civilization with political traditions far removed from those of the Arab tribesmen who came to replace Byzantine and Sassanid rule. The new rulers were themselves riven by bitter and bloody disputes about the character of legitimate rule in the Muslim polity. In the circumstances, the principle of election had no practical significance, nor was there any way by which the ruled could ensure that the ruler's actions conformed to the *shari'a*. Of the first four caliphs, the Well-Guided Ones, as they are known to Muslim tradition, the last three were assassinated, and the Muslim polity fell under the control of the Umayyad dynasty, established in Damascus in 661 CE. They in turn were forcibly overthrown some ninety years later by the Abbasids, who founded Baghdad and made it their capital. The hereditary rulers of this extensive empire (whether Umayyad or Abbasid) were attracted to notions of kingship by divine right. On such a theory, as it had been developed by the Byzantines and Sassanids long before Islam, the state is a divine ordinance which is personified by the king. The king is directly chosen by, and responsible to, God alone. He stands between God and the people, and ensures stability by maintaining both in their proper places. He is, to use a title which came to be officially given to Muslim rulers, the Shadow of God on Earth. Between this theory and the jurists' view of government the gap is immense. How early in Muslim history this gap appeared may be appreciated, for example, from the political testament of al-Mansur, the second Abbasid caliph, to his son, in which he identified the authority of the caliph with God's authority. And Mansur was merely continuing the tradition of the Umayyads. If the contrast had remained as absolute as it was at the outset, the jurists' theory might have remained as a foil to the other theory, and might have served as a challenge and as a rebuke to the successive despotisms which the world of Islam had to endure. But in fact, the jurists themselves came to modify their position significantly, in order to come to terms with reality, limit the damage to religious values, perhaps tame somewhat the violence of despotic rule, and make it bearable to the faithful.

How the Muslim divines came to approach politics may be seen from the argument in the continuation of the passage from Ghazali quoted above:

that the world and the security of life and possessions cannot be organized except by having a ruler to whom obedience is given, this is confirmed when we observe the civil wars which break out at the death of rulers and imams, especially when there is a lapse of time before another ruler has been nominated. Disorder, violence, and scarcity rule, cattle perish, and manufactures die out. Might seizes what it can; no one devotes himself to worship or to learning, even if he remains alive, while the majority perish by the sword. This is why religion and rule have been called twins. It is said that religion is a foundation and rule a guardian; what has no foundation is destroyed, and what has no guardian is lost. On the whole, no sensible man will doubt that human beings, with all their classes and varieties of desires and opinions, if they were left to their condition, and if there had been no one opinion to command obedience and to unify them, would have perished altogether. This is a disease which has only one cure: a strong ruler to command obedience and unify the diversity of opinions.

Here is a sober, indeed a bleak, account of the predicament of Muslim society and of what, at best, can be expected from a ruler. The prospect of disorder and anarchy is so dismaying that it has to be banished at all costs, and stability becomes the supreme political value. As Ghazali also wrote: 'The tyranny of a sultan for a hundred years causes less damage than one year's tyranny exerted by the subjects against each other.'[4] Here then is a point of contact and affinity between Muslim political thought and earlier theories of kingship by divine right. In one case, of course, stability meant a hierarchically ordered society, which was maintained and guaranteed by the divinely appointed ruler, while in the other, stability was a boon devoutly wished for by a society subject to lawlessness, rapine, and violent death. This intellectual affinity is demonstrated in the passage from Ghazali just quoted. Religion and rule, Ghazali writes, have been called twins: this maxim is of Sassanid origin; but however it was meant or understood by its originators, it is echoed and re-echoed in Muslim writings in order to emphasize that a strong ruler, whatever the origin of his power, whoever he is, whether he is bad or mad, is necessary to safeguard the *umma*. Again, if the preservation of the *umma* is the highest political value, then the expression of individual opinion which might lead to dissension and the disruption of communal solidarity is a crime to be put down with the utmost severity.

In attempting to deal with the vicissitudes of Muslim politics, the jurists made passive obedience to the holder of effective power a religious duty. They were perfectly well aware that this was a far cry from what the Law required, namely that a caliph, properly speaking, had to be a member of Quraysh, the Prophet's tribe, that he had to be physically and morally sane, to have knowledge of the *shariʿa*, and exercise his authority in accordance with it. Thus Ghazali, discussing the circumstances in which an imam may be deposed, lays down the condition that deposition must not lead to dissension or violent commotion: 'If this is impossible, then obedience is due to him, and his imamate must be accepted.' And Ghazali justifies his conclusion thus:

The concessions made by us are not spontaneous, but necessity makes lawful what is forbidden. We know it is not allowed to feed on a dead animal [i.e. one which has not been ritually slaughtered]: still it would be worse to die of hunger. Of those who contend that the caliphate is dead for ever and irreplaceable, we should like to ask: which is to be preferred, anarchy and the stoppage of social life for lack of a properly constituted authority, or acknowledgement of the existing power, whatever it be? Of these two alternatives, the jurist cannot but choose the latter.

Thus, starting from a position where all politics was religious politics, the jurists, in order to secure a modicum of protection for the religious life, paradoxically reached a view of politics in which any arbitrary power, provided it was successful, became legitimately entitled to the obedience of the subjects. As a maxim formulated by Moroccan jurists succinctly put: 'To him who holds power obedience is due.' This view, inculcated over the centuries by religious teaching and bitter practical experience alike, sums up the dominant attitude to politics in the world of Islam. Encapsulated in it, though never fully worked out, is a view of politics as an autonomous activity, and also a (necessary) distinction between politics as a public activity on the one hand, and the personal character and behaviour of the holder of public office. The distinction is to be seen in the jurists' injunction that jihad was to be pursued alongside all imams, whatever their conduct, that imams were entitled to levy the taxes legally due whatever the use to which they put them, that Friday prayers and other public religious rites ought to be performed with those who exercised rule, whatever their conduct or reputation. The sixteenth-century jurist, Sheikh Ibrahim al-Halabi, whose views were officially considered as authoritative in the Ottoman Empire, sums up this teaching as it was received at the threshold of modern times. 'The dignity of the imamate', he wrote, 'does not absolutely demand that the imam be just, virtuous, or irreproachable, or that he be the most eminent and the most excellent of the human beings of his time'; and again: 'Vices or tyranny in an imam do not demand his deposition'.

This sober, realistic, and conservative view of politics as a broken-backed but necessary activity, and of *raison d'état* as inescapably intermixed with it, is one to be found equally among all groups in Islam. But the Sunni jurists add two other points. The imam, they declare, is not infallible, and the imam must not be concealed. The Sunnis form the overwhelming majority of Muslims. They consider themselves the upholders of orthodoxy, that is of an authentic catholicity deriving from the sayings and doings—the *sunna*—of the Prophet himself. The Sunnis define their stand in opposition to those whom they consider to be narrow and divisive sectarians, principally the followers of Ali, commonly known as Shiites. These partisans of Ali, *shi'at 'Ali*, quarrelled with the main body of the faithful over the issue of the legitimate ruler of the Muslims. Shiites believe that Ali, the cousin and son-in-law of the Prophet, together with his direct descendants, are the only legitimate imams in Islam. Ali, the Shiites hold, was specifically designated by the Prophet as his successor, and his descendants also were each so designated. These imams enjoyed not only sole legitimacy, but also infallibility. The Twelver Shiites, today the main body of Shiites in the world, further believe that the twelfth imam in the Alid succession withdrew himself from the world in 874 CE, that he is now alive, and that he will, in his own good time, reappear again as the Mahdi—the Divinely Guided One—to re-establish forever the reign of justice in the world. The Twelver Shiite cause for long enjoyed no political success and, until the sixteenth century of the Christian era when the Safavids made Shiism the state religion in Persia, it had no territorial base. But this doctrine of a superhuman Hidden Imam as the only legitimate and infallible ruler of Muslims was clearly subversive of established authorities. Sunni rulers at various points found themselves endangered by a potent and tentacular propaganda using the notion of an infallible and Hidden Imam to threaten their position. The threat to the Abbasid caliphate proved at one point formidable, indeed nearly lethal. In 909 CE, a certain Ubaidullah, claiming descent from Ja'far al-Sadiq, the imam sixth in succession to Ali, and from his son Ismail (himself believed to have been the Mahdi), proclaimed himself Mahdi and Commander of the

Faithful in North Africa. The Fatimids (as his dynasty was called) eventually founded a formidable empire with Cairo as its capital, which lasted until 1171. Its rulers claimed to be such by the designation of their predecessors, and also to enjoy infallibility. At the height of their power they nearly succeeded in bringing down the Abbasid caliphate, for the threat they posed was not only military, but also ideological. The propaganda they conducted over the whole Muslim world aimed at discrediting the legitimacy of the caliphate unless the office was occupied by that descendant of Ali and his wife Fatima whose succession was divinely predetermined from Adam onwards, who was designated by his predecessor, and who was immune from error and sin. It is thus not surprising to see the Sunni jurists insisting that the imam, i.e. the legitimate ruler of the Muslims, must not be concealed and must not claim to be infallible.

But whatever the fears conjured up among Sunni rulers by the idea of the Hidden Imam, the fact remains that Shiite jurists inculcated in their followers the same doctrine and attitude of passive obedience which obtained among Sunnis. As a dissident group dwelling among hostile Sunnis the Shiites were taught by their jurists to practise *taqiyya*, i.e. to dissimulate their true belief for self-preservation. The Imam was not expected to revolt against the existing government, illegal as it was, and rebellion without his authority was unlawful. This was laid down by Hisham ibn al-Hakam who formulated the basic doctrines of the Shiites and who died in 795–6, i.e. before the occultation of the Twelfth Imam. The same teaching is repeated thereafter. Al-Sheikh al-Mufid, who died in 1022, taught that only the Twelfth Imam on his reappearance could lead a revolt; his predecessors had disapproved of any incitement to revolt, and enjoined 'dissimulation, restraining the hand, guarding the tongue, carrying out the prescribed worship, and serving God exclusively by good works'.[5]

The achievement of the jurists, then, was to discern with a clear-sighted shrewdness the grim political condition under which Muslims had to live; and to fashion, and inculcate, a reasoned, coherent attitude to this lamentable state of affairs.

Religion, they hoped, would thus be shielded to the utmost, and a society would survive in which Islam remained a public profession protected by the state, its rites regularly performed, and its teachings transmitted from generation to generation.

The final shape which Islamic political doctrine took—and it has remained the same to the present day—was a far cry from the simple and straightforward views of the early Muslims. These constituted a small religious polity highly coloured by the tribal ethics and customs of the Arabs who were Muhammad's first followers. But tribal political institutions, and the simple arrangements for choosing heads for the Muslim community in Medina and elsewhere, could not possibly cope with the administration of the newly conquered lands, which were extensive and closely settled by a non-Muslim population far removed in habits, beliefs, and sophistication from the new masters—primitive tribesmen garrisoning the land and living off its fat. The inadequacy of early Arab Muslim political institutions is not surprising, for tribal government is primitive government. Such a description is not meant pejoratively, but simply to indicate that tribal government is not an intricate or a tough kind of organization. As has been seen again and again, it cannot withstand contact with more solidly organized, and more sophisticated, political and legal structures.

But whereas tribal or primitive government is generally destroyed or emasculated by a foreign conqueror, the original features of the Arab Muslim polity were utterly changed and emptied of significance by the Arab Muslim rulers themselves. These rulers very speedily transformed an unsophisticated tribal polity into one of the most sophisticated and most durable kinds of rule, that of oriental despotism, the methods and traditions of which have survived in the Muslim world to the present day. What the Muslim jurists did was to articulate and theorize the conditions of political life in oriental despotism, and to teach that it was compatible with a Muslim way of life.

In Karl Wittfogel's classic formulation, oriental despotism is a political arrangement in which the state is stronger than society.[6] The political sociology of oriental despotism is extremely simple: only two social groups can be identified, namely those who rule, and those who are ruled. There is a population which works, produces, and pays taxes; and there is what the historian A. H. Lybyer, writing about Ottoman government in the sixteenth century at the time of Suleiman the Magnificent, has called the ruling institution.[7] This is quite separate from the population at large. Its main occupation is warfare. To this end it ensures law and order among its subjects, and extracts the largest possible proportion of the wealth they produce in order to maintain an opulent court and an efficient army which can be used for territorial expansion. The Damascus *qadi*, Ibn Jamaʻa (1241–1333) echoing a famous analogy which Muslim writers on politics had made—an analogy which may be much older than Islam—theorized this state of affairs by writing:

The world is a garden, the fence of which is the dynasty. The dynasty is authority supported by the army. The army are soldiers who are assembled by money. Money is sustenance brought together by the subjects. The subjects are servants who are reared by justice.[8]

Marx's view of the relation between economic and political power is here diametrically reversed. In oriental despotism, it is not ownership of the means of production that determines who will rule; rather, possession of military and political power determines who will enjoy the fruits of labour. In oriental despotism, then, economic power, properly speaking, is non-existent, riches are precarious, and property has no security. The story told of a Chinese general could never have been told of a West European magnate. The general was accused of plotting against the emperor. In order to exculpate himself and prove that he had absolutely no interest in politics, he claimed that his one consuming passion was to accumulate property in land. The insignificance and precariousness of private property under oriental despotism, where wealth can be confiscated or bestowed at the caprice of the ruler, and where to be suspected of being wealthy can invite sudden and utter ruin, may also be illustrated by an instructive and moving story told by Ibn Khaldun. The philosopher Ibn Baja happened to recite at an Andalusian court a poem which made the ruler so enthusiastic that he vowed that Ibn Baja should go home walking on gold. Ibn Baja found himself in a predicament: to refuse or to accept the gift was equally dangerous. To refuse would offend and anger, while to accept would make Ibn Baja so rich that his life might henceforth be at risk. Ibn Baja resolved the dilemma by begging for two pieces of gold which he proceeded to insert one in each shoe, and so walked home literally on gold, thus making the ruler's vow come true, and yet preserving himself from the perils of great wealth.

The absolute dominance of the ruling institution in oriental despotism, and the amalgamation in the ruler's person of both political and religious authority, is greatly in contrast with the manner in which political and religious authority came to be disposed in Western Europe in the centuries immediately following the decline and fall of the Roman Empire. During this period, *regnum* and *sacerdotium* were progressively distinguished one from the other. Feudal institutions also took shape then which led to the dispersal of military power and political authority among territorial magnates. These magnates, and the king who stood at the apex of the feudal order, were bound to one another and to their followers and dependants by an intricate and enduring web of rights and obligations which may be seen to be at the origin of the checks and balances which characterize modern constitutional government.

By contrast, in oriental despotism, no interest in society can resist the demands of the ruling institution or withstand its power. It is in this sense that in oriental despotism the state is stronger than society. Various attempts have been made to explain how this kind of rule came to take root and endure. But whatever the cogency and persuasiveness of these attempts, what is not in doubt is that extensive areas in Asia have remained for centuries under its sway, and that the caliphate became the legatee of oriental despotism. That it did absorb the methods of oriental despotism, and successfully adopt its way of controlling the subject, is not simply a matter of primitive conquerors being attracted and dazzled by a sophisticated and powerful foreign political tradition—of Persia, so to speak, making Arabia captive. It is rather that the very success of the Arab assault on the Byzantines and the Sassanids; the rule over large numbers of non-Muslims which was its consequence; the jealousies, the internecine struggles, and the vast cupidities which this unheard-of triumph conjured up, all served to render inefficacious and unworkable the political arrangements of the early Muslims, and led inevitably to the adoption of a system such as oriental despotism.

This development is seen fully manifest as early as the Umayyads. The caliph is the deputy of God Himself. He is the Shadow of God on earth. He sits in his palace quarters in formidable and solitary eminence, hidden from the gaze of his subjects, the capricious author of their good fortune, or their misfortunes. In the twinkling of an eye, he can raise the beggar to become his vizier and confidant, and he can suddenly and for no apparent reason humble and destroy the most powerful of his favourites. This is what the Abbasid Caliph, Harun al-Rashid, did to his viziers the Barmecides. The emblem of this terrible power is the black executioner who, in *The Thousand and One Nights*, is shown to be in constant attendance on Harun al-Rashid. Nearness to the supreme power is perilous. The constant care of the ordinary subject is to avoid the attention of authority. A story in *The Arabian Nights* concerns a householder who, coming back from work in the evening, finds a corpse near his door. He is terrified to report his discovery to the police, lest they accuse him of murder. He therefore carries the corpse and puts it over the wall into a neighbour's garden. The neighbour in his turn discovers the body and is likewise terrified. He in turn transfers it to his neighbour's property where the same story is repeated, and so on until the body has gone round the whole town without anybody daring to make its existence public. A saying attributed to the Caliph al-Ma'mun—Harun al-Rashid's son—declares that 'the best life has he who has an ample house, a beautiful wife, and sufficient means, and who does not know us and whom we do not know'.

Mutual mistrust thus isolates government and subjects from one another. But mutual mistrust and fear are at their height in relations between the ruler and his immediate servants. Mirrors of Princes and similar writings abound in advice to rulers and to those in their proximity about the best means of protecting themselves from the dangers they are likely to meet. Rulers are advised to take precautions against poisoning, to set spies on their ministers, to beware of close friends, wives, brothers, and progeny. Viziers are warned to mistrust fair words when uttered by their masters, since they are usually meant to produce false security, and to be a prelude to the minister's downfall.

A usual concomitant of this separation between the concerns of the rulers and those of the ruled is the prevalence of what is called corruption. Officials are servants of the ruler who, at his will, appoints them to exercise power over the subjects, and at his pleasure deprives them of this power. So long as they continue to enjoy the ruler's countenance, officials must be propitiated by the subjects who have to seek their favour or avert their ill will. In this atmosphere of fear, mistrust, and caprice, officials in turn have to propitiate their superiors, and these superiors in their turn their own superiors. The ruler himself who bestows offices, but sometimes also sells them, thus came to be enmeshed in a vast network of bribery, along with his servants and subjects. We may even say that oriental despotism and bribery are inseparable, that bribery is a necessity in this kind of regime. It is a necessity for the official who desires to enjoy to the utmost his tenure while it lasts, to compensate himself for the sum he would often have had to expend for the purchase of his office, and to make provision against sudden fall from favour. It is a necessity likewise for the subject, since only bribery (or personal favour) can secure him a lodgement, more or less precarious, in the interstices of power.

The separation between the ruling institution and its subjects is seen at its most perfect in the Mameluke regime which controlled Egypt and the Levant from 1230, and in the Ottoman Empire which destroyed this regime in 1517. The Mamelukes were white slaves, usually Turkish or Circassian, whom Muslim rulers bought for use as soldiers. The Ayyubid sultans in Egypt were among the rulers who made use of these slave-soldiers. But the soldiers supplanted the dynasty and themselves became the rulers. Perpetually replenished by recruits of the same kind, isolated by language, avocation, and status from the populations they controlled, the Mamelukes disposed of the fate of Egypt and other territories under their sway solely through the internecine intrigues and armed clashes in which their various factions were perpetually engaged. The ruling institution of the Ottoman Empire was likewise utterly cut off from the population at large. Here the sultan was not himself a slave-soldier but a descendant of the Ottoman dynasty. He, however, controlled an official establishment the members of which were slaves acquired through the *devshirme*, that is the periodical collection of male children from his Christian subjects—children to be subsequently trained for either a military or a civil career. The character of this establishment is well described in a fifteenth-century work by a close European observer—when the Ottoman system was still in its vigour—whom Lybyer quotes in his seminal work, *The Government of the Ottoman Empire at the Time of Suleiman the Magnificent*. Out of the slaves acquired by the *devshirme*, this report declared,

promotions are made to the offices of the kingdom according to the virtues found in them. Whence it comes about that all the magnates and princes of the whole kingdom are as it were officials made by the king, and not lords or possessors; and as a consequence he is the sole lord and possessor, and the lawful dispenser, distributor and governor of the whole kingdom; the others are only executors, officials, and administrators according to his will and command . . . Whence it follows that in his kingdom, although there is an innumerable multitude, no contradiction or opposition can arise; but united as one man in all respects and for all purposes, they look to his command alone, they obey, and serve unwearingly.

It thus follows that in this kind of rule no institutions or regular arrangements exist whereby the ruled can convey their views or wishes to the rulers, let alone ensure that such views are taken into consideration. Not only is the idea of representation, or of elections, unknown to Muslim political thought or practice, but Muslim rulers looked with suspicion on any autonomous tendencies evinced by the subjects in setting up bodies like trade guilds, and tried systematically to control their activities. It is also significant that municipal institutions and corporate bodies in general are unknown in Islamic polities. Figures like the 'omda or the *mukhtar*, the headman of a village or a city quarter, are by no means representatives; rather, they are appointed or approved by the government and responsible to it alone for the good order of their districts.

Besides the fundamental cleavage between rulers and ruled, other cleavages characterize traditional Muslim society. A significant one is that between town and country. From the start Islam has looked down on the countryside, on the fellah's way of life and his pursuits. The Prophet is reported in a *ḥadith* as remarking of a plough that 'This thing does not go into a house without making it lowly'. The fellah is lowly for another reason: unlike the bedouins who constituted the first armies of Islam, and whose ethos took on all the prestige which accrues to conquerors, fellahin are attached to a specific locality and identified with a particular piece of land; they have no record of their lineage and preserve no account of the valiant deeds of their ancestors. In the world of Islam, the town is the centre of power, where the ruler, his court, and his officials live, while the countryside exists in order to be milked for the benefit of the ruling institution and its hangers-on. Hence, the absence generally in the Muslim world of a landed gentry who in Western countries have developed hereditary local influence and a local patriotism and who, taking their due place in an ordered polity, yet have a feeling of sturdy independence, do not form part of an official, salaried hierarchy, or have to curry favour with a royal court, or feel beholden to it. In the Muslim world, it is only of nomadic tribes that some of these attributes can be postulated. But these tribes, unlike the landed gentry of the West, were not incorporated in the Muslim polity. Rather, they remained outside, a disturbing and destructive element, and, in a process vividly described by Ibn Khaldun, occasionally succeeded in overwhelming an established dynasty, but then took over from their victims the very same old centralizing and despotic traditions.

From the start, Muslim society incorporated non-Muslim communities. As the Koran shows, the Prophet himself was acquainted with Judaism and Christianity and the Muslim claim is indeed that he is the seal of the prophets, i.e. that Islam completes and perfects the two earlier divine revelations. When Muhammad and the small band of his original followers retired from Mecca in order to escape opposition and persecution, they went to Yathrib, which came to be known thereafter as *madinat al-nabi*, The Prophet's city (Medina in English usage), whose three Jewish tribes had long been established. Though relations between them and the Muslims began by being cordial, Muhammad soon clashed with the Jews, who were defeated in battle, and of whom two tribes were expelled, while the menfolk of the third were decimated and their women and children sold into slavery. Subsequently, other Jewish and Christian settlements in Arabia entered into pacts with Muhammad whereby they offered submission and paid tribute. The Koran sets out the manner in which followers of Scriptural religions were to be treated:

Fight against those who do not believe in Allah nor in the Last Day, who do not forbid what Allah and His Apostle have forbidden, nor practise the true religion, among those who have been given the Book, until they pay the *jizya* from their hand, they being humbled.[9]

Jizya is the tribute, later understood to be a poll tax, which Christians and Jews had to pay in humility and in token of submission to the Muslim ruler. In return they received protection for good behaviour. In the Muslim polity these non-Muslims were allowed by the ruler to organize their own communal affairs under their own leaders who enjoyed wide judicial and administrative powers in respect of their communities, each of which was left free to follow its own (religious) law in matters of personal status and of inheritance. This is the well-known *millet* system (from the Arabic *milla*, community) which was maintained by successive Muslim rulers in the Middle East, and which survived—albeit in an attenuated form—into the twentieth century. Non-Muslims in the Muslim polity do not therefore form part of the Muslim *umma*. To use a European distinction, they do not belong to the *pays politique*, but only to the *pays légal*. So long as they accept their inferior status, and do not pose a threat to the security of the *umma*, they are entitled to protection and are allowed a degree of what might be called communal autonomy. This state of affairs is strikingly different from that obtaining in medieval or early-modern Europe. Here, the pressure for religious uniformity was constant and unremitting. The Jews were practically the only non-Christian element in Europe, and Jewish communities by no means enjoyed the recognized status which was vouchsafed to them and to Christians under Islam. The existence of these non-Muslim communities, each inhabiting its own quarter, and like the Muslims moved, not by local patriotism or identification with a particular territory, but exclusively by communal solidarity, enhances yet further the impression of a segmented society, all gaps and separations: separation between ruler and ruled, between town and country, between the nomad and the settled population, between religious communities.

This segmentation meant that the reach of the oriental despotism which controlled Muslim society did not go as far as one might think. Traditional Middle Eastern rulers simply took no interest in a vast range of issues in which modern governments of all colours assume it their right and duty to intervene. Middle Eastern governments, for instance, did not feel obliged to make provision for the education or welfare of their subjects. Again, it did not occur to them that they could mobilize their subjects, and canalize their energies in support of their aims and policies. In any case, they did not have the technical means to indoctrinate and control large masses of people. In the absence of modern means of communication, and of an abundant supply of paper or of computers it was also impossible to maintain detailed supervision over the activity of office-holders and their dealings with subjects. Also, just as non-Muslims were left a large measure of freedom in their communal affairs, so the ruler took it for granted that certain affairs relating to the *umma* were the subject of divine prescription with which no one could tamper and were to be regulated by *qadi*s, or religious judges, according to the *shari'a*. All these, then, were built-in limitations on the activities of government under oriental despotism. These limitations left a great deal of elbow-room—no doubt politically insignificant—to the subject, and meant that there was a boundary, implicitly recognized on all sides, which divided the public realm from the private, which latter comprised a large variety of social activities.

2

The Modern World:
Threat and Predicament

It is generally agreed that decline overtook Middle Eastern society in modern times. Decline is a relative notion. In this case, what is meant is that Western European society was gradually becoming richer, more technically advanced, and militarily more powerful from around the seventeenth century onwards, leaving the Muslim world far behind. By decline it is also meant that economic activity and political and military institutions fell into decay, compared to their earlier condition, and there is no doubt that in the seventeenth century the Middle East was a poorer and more disorderly place than it had been a century or two earlier. Commerce and economic activity had shifted westward. The great inflation which the new, enormous supplies of gold and silver from the New World brought, played havoc with Ottoman finances and foreign trade, and created stress and discontent in the provinces. The mainstay of Ottoman power outside Constantinople was a network of freeborn soldiers, who, in a large number of provinces, in return for revenues from agricultural estates assigned to them by the Sultan, carried responsibility for order in their district, and when called upon provided levies of cavalry to take part in military campaigns. These estates were carefully graded according to the revenues they produced and were assigned to a particular beneficiary according to his rank and responsibilities. The estates were in no sense the property of the official or officer who enjoyed their revenue. He continued to enjoy the revenue while he remained in office; he could not bequeath the post to his descendants, neither, *a fortiori*, was he able to bequeath the property itself. These conditions approximated to the earliest character of European feudalism. And as in European feudalism, the enjoyment of the revenue which was originally tied to the discharge of an office, came to be the endowment of the office-holder in perpetuity: he was able to bequeath it to his descendants; and not only the revenue, but also the property itself which produced the revenue. Likewise tax-farmers who undertook to collect the tax and to remit an agreed yearly amount to the central government in due course became owners of the land whose produce they taxed. This of course was an abuse of the system. But it was an abuse which led to the appearance of a class of landed notables who disposed of an armed following and of hereditary wealth, and who could thus act as a counterpoise to, and a check on, the despotic power of the central authority.

Other Ottoman institutions could also be said to have decayed and declined; but, by the same token, in declining from their original state, they might be considered to offer new possibilities and a new promise to the body politic. The Janizary corps was the most important component of the Ottoman standing army, recruited by means of a levy of Christian children from Ottoman provinces, the *devshirme*. It was for long a most formidable organization. Always on an active footing, its members—slaves of the sultan—lived in barracks and were forbidden to marry and have families. In due course corruption overtook this organization. Its privileges were valuable, and freeborn Muslims, in no sense soldiers, succeeded in having themselves inscribed on its rolls and made eligible to receive the appropriate wages. To be a Janizary in the end became a valuable hereditary perquisite; the corps became useless for military purposes, but it remained a formidable, if disorderly and anarchic, force brought to bear on public issues in the capital. The weakening of the central power also meant that provincial governors in the Balkans, in the Levant, in Mosul, Mesopotamia, and elsewhere would arrogate power to themselves and render merely formal obeisance to the Sultan. And in Egypt, the Mamelukes, whose power was broken in 1517, again became the absolute rulers in all but name—a threat and an embarrassment to the Sultan. For him, for the

Ottoman dynasty, this was undoubtedly decline. But from this decline possibilities of a new political order might have arisen, in which checks and balances, and a limited government, would become the norm, and would not be simply the fortuitous and temporary outcome of the vicissitudes afflicting the old despotic regime.

The affair of the *sened-i ittifak*, i.e. the deed of agreement, of 1808 serves to suggest what might have been. In the previous year Selim III had been deposed by the Janizaries who were afraid of the military reforms on which the Sultan had embarked. But in 1808 the Janizaries' Sultan, Mustafa IV, was deposed and Selim's cousin was proclaimed Sultan as Mahmud II. The kingmakers were the provincial notables who had marched to Istanbul with their own armies to overawe the Janizaries and the court. These notables made Mahmud sign a deed of agreement which was a regular contract according to the form recognized by law. One party to the contract was the Sultan, while the other party was constituted by the representatives of the notables. These acknowledged the supreme authority of the Sultan and his right to levy taxes and to conscript. In return the government promised to respect the status and established rights of the notables. These were guaranteed in the possession of their land and confirmed in their right to bequeath it to their heirs, who were also bound by the agreement. The notables also bound themselves to respect these rights among themselves, and as they related to the lesser notables under their jurisdiction. If officials or the Grand Vizier himself violated the agreement or acted corruptly, the signatories would stand forth as accusers and secure removal of the abuses. The Sultan was to levy taxes justly and fairly. The notables also undertook to march on the capital in order to uphold the government should they come to hear of any threat to its existence.

To compare this document to Magna Carta is natural, but whereas Magna Carta reflected the balance of power between king and barons and articulated feudal realities and feudal traditions, the *sened* did nothing of the kind, and speedily proved to be a dead letter. And this was not simply, or principally, because the deed of agreement lacked institutional embodiment or the means to enforce its provisions. Rather it was because new initiatives taken by Ottoman and other Middle Eastern rulers were giving a new lease of life to oriental despotism, and rendering entirely illusory the prospect of constitutional or limited government. Selim III was deposed in 1807 because the Janizaries and their supporters and sympathizers objected to the military reforms which, after his accession in 1789, Selim had made a sustained attempt to carry out. The Sultan was led to undertake these reforms in the conviction that the state was in mortal peril from its European foes who had the advantage of superior weapons and a much better military organization. This advantage had meant a series of defeats for the Ottoman Empire, beginning with the retreat from Hungary and Transylvania, acquired in the sixteenth century, but ceded to Austria under the terms of the Treaty of Karlowitz of 1699. During the eighteenth century, other losses continued to be incurred in Europe. These culminated in the cession of substantial territories to Russia by the Treaties of Kuchuk Kainarja (1774) and Jassy (1794)—Treaties which followed, and ratified, serious military defeats. But in spite of what Selim, his Ottoman successors, and other Middle Eastern rulers did, loss of territory, helplessness, and political inferiority continued to afflict them throughout the nineteenth century and the first half of the twentieth century. Egypt was invaded and occupied for a few years by the French at the end of the eighteenth century. The French also invaded Algiers in 1830 and annexed it; they imposed a protectorate on Tunisia in 1881. Egypt was occupied by Great Britain in 1882. The British established their supremacy in the Red Sea, the Persian Gulf, and Southern Persia by the first decade of the twentieth century; Italy invaded and occupied Tripoli and Cyrenaica in 1911. Russia exercised unremitting pressure against Ottoman and Persian dominions all through the nineteenth century: it annexed the Caucasus, and the Central Asian emirate of Khokand; established a protectorate over the two neighbouring emirates of Khiva and Bokhara; and succeeded, in the first decade of the twentieth century, in having its paramount position recognized in northern Persia. Military weakness and Great-Power pressure led the Ottomans to concede autonomy and independence to subject populations in Greece, Serbia, Romania, and Bulgaria, and some of these new states would actually attack, and imperil the existence of the Ottoman state in 1912–13. The First World War saw the defeat and dismemberment of the Ottoman Empire, and the control of its Arabic-speaking territories pass to Great Britain and France. Finally, the Jews of Palestine, in 1948 and subsequently, were powerful enough to defy Arab opposition to the establishment of Israel, and to defeat Arab forces in successive encounters. This long, unrelieved series of defeats, set-backs, and retreats, so unlike what Muslims had experienced throughout their history, and at such variance with their view of themselves and their relation to the non-Muslim world, has to be kept continually in mind, since it is the implicit and frequently unspoken assumption underlying political action in the Middle East in modern times. In all countries which had been non-Muslim and which were conquered for Islam, the preacher of the Friday sermon holds a wooden sword in his hand. The Friday sermon is the ruler's prerogative which he delegates to the preacher, and the sword in his hand is a striking symbol and reminder for the congregation of the power and sway of Islam. In the modern world there has come to be a progressive disjunction between the symbol and what it is meant to symbolize, while the reminder points up the bitter contrast between past glories and present humiliation.

Over a period of some fifteen years, Selim III tried to set up a new model army which would combine the organization and discipline characterizing modern armies with proficiency in the use of modern weapons. But two entirely different armed forces could not subsist side by side, and as has been seen, the Janizaries overthrew Selim and attempted to undo his innovations. They seemed to be successful, and, although the Sultan they put on the throne was quickly overthrown and replaced by Selim's cousin, Mahmud II, the new Sultan himself owed his position to the provincial notables who had marched on Constantinople. If the *sened* which he was then compelled to sign were to be effective, then his scope for taking up the cause of reform would have been so narrow as to be virtually non-existent. But the new Sultan worked to ensure that the *sened* would remain a mere piece of paper, and no more.

Mahmud's reign was a long one, lasting until his death in 1839. It was, however, only after he had ruled for eighteen years that he was finally able to deal a death-blow to the Janizaries and other traditional military formations, and seriously to set about forming a European-type army. But in this endeavour he had been preceded by another powerful and dominating figure, Muhammad Ali, the Pasha of Egypt. Muhammad Ali had arrived in Egypt as an officer in the Ottoman forces sent to Egypt following its occupation by Bonaparte. By 1805, he had managed to eliminate rivals and to be recognized by the Sultan as governor. Thereafter, he set about destroying all centres of power in the country which might threaten his unfettered rule or pretend to some independence of action. He tackled the Mamelukes who had succeeded in regaining the power which had been theirs before the Ottoman conquest, and who, before the French invasion, were the effective rulers of Egypt. This invasion had seriously weakened them, but they could still act as a check on the governor, as the Janizaries could on the Sultan in Istanbul. Muhammad Ali fought the Mamelukes and destroyed their military power. Finally, in a spectacular act of treachery with which his memory has remained popularly associated, he assembled them for a banquet at the Cairo citadel and carried out a mass assassination.

With the help of foreign officers Muhammad Ali set about creating a new model army, drilled and disciplined in the European manner. At first he tried to man this army with slaves obtained chiefly from the Sudan, which he invaded in 1820–1. But he found that these slaves made unsatisfactory soldiers and, following the advice of his foreign experts, he began to conscript Egyptian fellahin. But the fellahin were naturally unwilling to serve, and violent means had to be used for their coercion. Military modernization had thus meant a very great increase in the demands of the ruler over the subject. In order to expand his resources, and thus to be able to afford a powerful military and naval force, Muhammad Ali made himself wellnigh the sole landlord in Egypt, and the sole buyer, seller, and manufacturer, exporter and importer of a wide range of agricultural and other products. When he exterminated the Mamelukes he declared all their lands forfeit to himself. On various pretexts he also laid hands on extensive properties which over the centuries the piety of donors had put in trust for the upkeep of mosques, religious schools, and other charitable purposes. Muhammad Ali also increased greatly, by one means or another, the amount of tax to which the cultivator was liable, and likewise levied high taxes on those urban manufactures and activities which he did not directly control. Even prostitution did not escape the tax-gatherer's net. In short, as one of his subjects put it, Muhammad Ali was jealous of the very lice that eat up the fellah.

Not until the second half of the twentieth century did a Middle Eastern ruler aspire to control his subjects so systematically and so comprehensively. Certainly Muhammad Ali's Ottoman suzerain, Mahmud II, attempted nothing so ambitious and grandiose, but he too, having bided his time and prepared his ground, at one fell swoop in 1826 destroyed the Janizary corps and other traditional military formations, and substituted for them a European-style army dependent on conscription. Both before and after this revolution, Mahmud subdued the various notables and Pashas, both in the European and Asiatic provinces, who had taken advantage of the decay of the original Ottoman system to free themselves from the Sultan's control. If the Janizaries, the notables who controlled so much landed property and could deploy large numbers of armed retainers, the virtually independent Pashas such as those of Baghdad or Mosul, or the Baban dynasty in southern Kurdistan, were so many arbitrary and anarchic elements in the state, they also represented substantial interests which, in their fashion, actually or potentially served to counterbalance, and thus limit, the sway of oriental despotism. Again, the very extensive charitable endowments, regulated by the religious establishment and controlled by a multitude of trustees specifically designated in the various instruments of trust, also represented a substantial interest outside the ruler's control. As has been seen, Muhammad Ali, in one way or another, by force or subterfuge, laid his hands on many of these endowments in Egypt. Mahmud II likewise tried to do so. Following the destruction of the Janizaries in 1826 he established a government department which took over the administration of the endowed properties, and the upkeep of mosques, colleges, and the religious establishment in general.

The new European-model armies, then, meant an enormous increase in the ruler's power, and a concomitant increase in centralization. There were various reasons for this concentration of power and of administrative processes in the hands of the ruler, his officials, and the central departments of government. Some of these were now reorganized and, it was hoped, streamlined; others were created in response to new needs and new responsibilities. The new armies were based on conscription. Conscription required the keeping of detailed population records—records which had to be available to the central government. Large conscript armies had to be administered and trained in ways quite different from those appropriate to the Janizaries or similar formations, or to the local levies which in times of emergency supplemented the standing army. Even if the ruler had not desired to concentrate power in his hands—which he did, precisely in order to avert the external and internal threats which had made necessary the destruction of Janizaries or Mamelukes—the new kind of organization, which depended for its efficacy on meticulously working to a uniform rule, itself implied and entailed centralization. This kind of administrative work in its turn required the detailed supervision by those at the top of subordinates unfamiliar with new methods and thus all the more unwilling to take decisions on their own, even if these were simply routine. Administrative channels, hence, were always clogged, and administration fussy and top-heavy.

A manual of Ottoman public administration dating from the end of the nineteenth century describes the progress of an application to a government office from a member of the public. The letter of application cannot be entertained unless it bears a stamp of a prescribed value. Since paper is a perishable commodity, and documents have a way of going astray, the postulant seeks to place his request, duly stamped, directly in the hands of the departmental head or of his man. The document has first to be entered at the central registry. It is then sent to an official whose function it is to direct papers to the appropriate official for decision. The paper then goes back to the central registry where it is dispatched to this latter official. He in turn either delivers a ruling, or else decides that the matter has to be examined by yet another official. In this case, the papers resume their peregrination, but if a ruling has been made, then the file is sent to another official who is charged with writing out the decision in due and proper form. This document then goes to the registry where it is entered, and where the petitioner has to apply for it himself. It is obvious that the ordinary subject, unversed in this modern science of public administration, and most likely illiterate, will be easily lost in this labyrinth, and that occasions when bribes and benevolences must be applied have greatly multiplied. Administrative modernization was introduced not only in the Ottoman Empire and in Egypt but also in Tunis which, like Egypt, was a quasi-independent Ottoman province. A French observer commenting on the effects of the reforms (introduced in 1857–61), noted that previously a subject depended, in his dealings with authority, on two men only, the governor, the *qa'id*, of his province, and the judge, the *qadi*. The subject was, as this observer put it:

eaten by two men, but only by two. Today when almost all the powers of the *qa'id* and the *qadi* have passed to the new tribunals, he is eaten by all the members of these tribunals. He used to accept the decision of the local judge because it was a prompt decision, and only in grave issues would he appeal to the Sovereign. Today, the inhabitant of Gabes, which is 80 leagues from Tunis, who wants to appeal from a judgment rendered by the local tribunal, has to go to Tunis. And even after a journey so costly for him, he does not enjoy the privilege of personally rehearsing his grievances before the Bey. It is yet another committee issuing from the supreme Council . . . which will annul or confirm the first verdict.[1]

Over-administration thus became a tradition which has lasted to the present day. It was grafted on to the older tradition that officials were not the servants, but the masters, of the public. Hence administrative reform was inevitably accompanied by a pervasive petty corruption, the beneficiaries of which were the new tribe of pen-pushers conjured up by new or refurbished departments of state. The burden on the subject, we may think, cannot have been any less than under the old dispensation when (to use the expressive term which the French observer of the Tunisian reforms borrowed from local usage) the number of officials who 'ate' was much smaller, even though the bribe which could satisfy their appetite had perhaps to be appreciably larger. The order of magnitude involved here may be appreciated from the fact that at the end of the eighteenth century the number of civil officials in Istanbul was about 1,500, while a hundred years later it is estimated to have been between 50,000 and 100,000. Furthermore, working to a rule, which reform necessarily demanded, was quickly transformed by officials afraid of initiative into a rigid routine highly unresponsive to changing needs or unusual situations.

It is not only that the traditional functions of government now became the affair of a finicky and vexatious bureaucracy. It is also that the logic of reform acquired an impetus of its own, and led to the state assuming new responsibilities with a view of promoting the welfare of the people. The Ottoman Government wished to ameliorate the condition of its subjects, and thus attempted to improve the collection of taxes and the administration of justice. It also wanted to promote education and public works, and to give the subject a voice in public affairs. Provincial and district councils were thus set up, beginning in the fifth decade of the nineteenth century. They were given extensive financial, administrative, and judicial powers which operated as a check on the governor and other officials who represented the central government. These councils, however, in no sense represented the population at large or

defended their interests. The councillors themselves belonged
to the official classes. Their connections and their education
enabled them to exploit to their advantage the opportunities
afforded by their official position, by the increased scope of
government, the greater complexity of administrative processes, and the proliferation of complicated official forms which
these processes required and generated. Provincial and local
government was reorganized more than once in order to
remedy the abuses which its creation inherently entailed. But
in the end, local councils proved to be yet another, largely
superfluous, layer of officialdom which the population had to
cope with. To this day, the reality of power has remained
tightly held in the hands of the central government and its
agents, while municipal or local government has remained
empty of substance or significance.

Reforming activity extended also to landownership and
tenure. By far the greatest part of the land belonged to the
state. As has been seen, it used to be granted in feudal tenure
in lieu of salaries to military and civil personnel. Rulers had
also appointed tax-farmers to collect the taxes due from various
districts. But abuses led to fief-holders and tax-farmers actually
becoming virtual owners of the lands whose revenues had been
assigned to them for a term, or whose tax they were simply
supposed to collect. Mahammad Ali confiscated the lands held
by the Mamelukes. Mahmud II likewise abolished feudal
tenures. Muhammad Ali and his successors then made extensive grants of land to themselves, their families, and officials.
The beneficiaries of these land grants also came to have the
right to bequeath them to their descendants. Subsequently,
from 1858 onwards, new laws made full private ownership of
land possible to the ordinary cultivator in Egypt. Extensive
land grants by Muhammad Ali are said to have been made with
a view to the creation of a landed aristocracy, while securing
the rights of the ordinary cultivator was meant to give as large
a proportion of the fellahin as possible a stake in the land. This,
too, seems to have been the purpose of the Ottoman Land Law
of 1858. If Ottoman agriculture was to develop and prosper as
did that of Western Europe, then the peasant must enjoy a

secure, unambiguous title to his property, and the laws governing land tenure must be made as simple and clear as possible.
Land was to be surveyed, ownership established, title registered, and title-deeds issued to the rightful owner. All this was
easier said than done, for the extensive bureaucratic organization needed for so large and ambitious an operation necessarily
fell prey to the same vices of dilatoriness, corruption, improper
influence, and arbitrariness which infected other modern or
modernized bureaucracies. Thus in a great many cases in very
many parts of the Empire, land was registered in the name not
of the cultivators who actually tilled it, but in that of tribal
chiefs and city notables who, with the help of the law and by
the magic of registration, acquired new and extensive riches.
The disorder in land tenure was by no means cured, but rather
made more complicated and intractable—a disorder which the
rulers of the successor states of the Ottoman Empire were to
inherit, grapple with, and often profit from.

But the political purpose of the land laws, either to provide
security for property, or to create a landed aristocracy, or an
independent peasantry with a stake in the land, proved a
failure. Nowhere in the Middle East did a landed aristocracy
come into being as a distinct political interest, aware of its own
position and independently minded. There were, of course,
large landowners, but landownership failed to be an element in
a (non-existent) system of political checks and balances. Landowners were simply rich men who might or might not personally
be members of the ruling institution. If they were, they would
be on a par with others who belonged to the official classes, but
who were far from being landowners. In Egypt, for instance,
the prominent political figures under Muhammad Ali's successors up to the abdication of King Faruq in 1952 did not derive
their position from the ownership of land. This is true of Riaz
and Sherif who were prominent ministers in the last quarter of
the nineteenth cenrury, as it is of Zaghlul, Nahhas, or Sidky
who occupied the political scene during much of the period
between the end of the First World War and the end of the
Egyptian monarchy. The provinces of Baghdad and Basra
(which, together with the province of Mosul, formed the

kingdom of Iraq after the First World War) are even more significant in showing not only that landed property does not lead to political power, but also that grants of land or of title to land by rulers leads to the political emasculation of the recipients. Baghdad and Basra contained many turbulent tribes whom the Ottomans had been long powerless to subdue. In 1869, Midhat Pasha, a prominent and successful Ottoman reformer, was appointed Vali of Baghdad. In his hands the 1858 Land Law became an instrument controlling powerful and anarchic tribal chiefs. Tracts which had been customarily used for grazing by a particular tribe, or on which cultivators belonging to a particular tribe had settled, and whose legal status was unclear, were now registered in the name of the tribal chief. This land became more valuable since Mesopotamia, now linked to the international market, was becoming an exporter of cereals on an increasing scale. The tribal chief who became a great landowner by virtue of the law could now enjoy undreamt-of riches, but he also became *ipso facto* a client of the state, dependent on its support should rival chiefs seek to contest his title. When the kingdom of Iraq was established, grants of title to land were used by the new rulers as a means of political control, securing support for a ramshackle regime which had no roots in the country, or rewarding armed rebellion undertaken in favour of various political factions engaged in internecine struggle, or else as a convenient and speedy way for political figures to acquire great wealth for themselves.

In his memoirs written after the collapse of the Iraqi monarchy in 1958, a minor political figure, Jamil al-Urfali, remarks that Prince Abd al-Ilah (who had been Regent for the infant King Faysal II from 1939 to 1953, and who thereafter remained very powerful as heir to the throne) was unwilling to anger the tribal chiefs by supporting the division of huge estates and their redistribution to small cultivators, because he considered these chieftains to be a shield against the military uprisings which periodically threatened the monarchy. But in fact, Urfali pointed out, the power of these chieftains collapsed with the collapse of the monarchy, 'since their wealth was simply derived from the state'. Property in land, however extensive, bestowed no power. It was itself highly insecure as was seen not only in Iraq, but also in Egypt, and in Syria where the Ottoman Land Law had, as in Iraq, made possible the creation of large landholdings. In all these cases property was neither dyke nor shield; its defences were rudely swept away by military regimes which met not the slightest resistance as they dispossessed, nationalized, and redistributed. As for the small cultivators, if they did benefit from land laws which sought to protect their interest—and most often the laws were circumvented or exploited to their profit by those with influence or with access to the magic world of legal and administrative procedures— then this did not result in their having a stake in the country or becoming a powerful and substantial political interest, able to make their voice heard in government, and render rulers responsive to their views. On the contrary, land redistribution meant that the peasant would become more subject to the demands of authority, and to the caprice, and possible oppression, of the bureaucrats introduced to supervise land reform, and administer the farmers' cooperatives which often accompany it. No one on the spot can really control the action of these officials. They are answerable through a long, complicated, often defective, and corrupt chain of command to a remote authority in the capital—an authority which has neither leisure to cope with the incessant flow of paper, nor incentive to right the paltry wrongs of obscure persons.

The setting up of institutions of local government, and the legislation on land ownership and tenure were all part of an ambitious policy which originated in military necessity, but which speedily assumed a momentum of its own. This policy is known in Ottoman history as the *tanzimat-i khayriye*, beneficent reforms. The official state of mind which made such a policy seem desirable was by no means confined to the Ottoman Empire; sooner or later it spread also during the latter half of the nineteenth century to Egypt, to Persia, to Tunis and, somewhat later, to Morocco. Ottoman developments may thus be considered to exemplify and to herald similar developments elsewhere in the Middle East. The state of mind which presided over the policy of reform was something quite new in Middle

Eastern politics. From its perspective, existing institutions and traditional arrangements, hopelessly outmoded, inefficient, and corrupt, call for change, more or less radical. When, however, a reform is effected, it inevitably becomes itself an object of dissatisfaction, and in turn has to be radically reformed. This state of mind may be summed up in the words used by the eminent Ottoman jurist and official, Jevdet Pasha (1822–95): *al-baraka fi'l-haraka*,[2] i.e. constant activity is a blessing. This dissatisfaction with what exists has marked the educated and official classes since the beginning of reform. That it continues to exist perhaps indicates that policies of reform, however laudable their aims and however crying the abuses they set out to remedy, have necessarily disappointed expectations. Attempts by governments to remedy abuses and to create a better society have brought in their train different—sometimes greater—evils and abuses. Whatever else they promoted, policies of reform, then, did have a cumulative propensity to promote restlessness and instability in modern Middle Eastern politics.

Two official documents are recognized as milestones in the history of the Ottoman *tanzimat*. They were both issued during the reign of Abdul Mejid who succeeded his father Mahmud II in 1839. The first, the Noble Rescript of the Rose Chamber, appeared during that year, while the second, the Imperial Rescript, appeared in 1856. The two documents are milestones in more than one sense. They were both published during a time of turbulence in the international relations of the Ottoman Empire. The first, the Rose Chamber Rescript, came when the Empire was in the throes of a crisis brought about by the defeat which the forces of the Pasha of Egypt, Muhammad Ali, inflicted on Ottoman arms at the battle of Nezib (near Aleppo) in June 1839. This defeat bade fair to disrupt the balance of power in the Middle East. It threatened not only the Ottoman regime, but the interests of various Great Powers, chief among which was Great Britain. Acting together with the Russians, the Austrians, and the Prussians, Palmerston compelled Muhammad Ali, who was supported by France, to evacuate the Levant, which he had conquered in 1831–2. British policy in the Middle East as it came to be formulated in the 1830s was based on the assumption that the independence and integrity of the Ottoman Empire was a British interest. The formulation of such a maxim of policy is itself an index of Ottoman weakness. The international repercussions which this weakness produced constituted the Eastern Question, which loomed so large in European diplomacy throughout the nineteenth century and until the outbreak of the First World War. To preserve the integrity and independence of the Ottoman Empire entailed, in British eyes, the reform of its public institutions. Since British protection was now vital for Ottoman survival, British influence at Constantinople became very great, and this influence was exercised in favour of reform. The Rose Chamber Rescript may be seen as an expression of Great-Power influence and pressure, leading to substantial changes in the internal affairs of the Empire.

This is even more true of the Imperial Rescript of 1856, also known as the Reform Decree, which was issued in the aftermath of the Crimean War. Great Britain and France fought and defeated Russia in order to prevent a Russian hegemony over the Ottoman Empire. The Treaty of Paris, which terminated the war, was signed in March 1856. In its ninth article it took cognizance of the Reform Decree which had been issued the previous month and formally communicated to the Peace Conference by the Ottoman representatives. The Reform Decree, therefore, was an earnest of the determination of the Ottoman Empire to change its institutions so as to become worthy of its admittance 'to participate in the advantages of the public law and system (*concert*) of Europe', as Article 7 of the Treaty of Paris put it. But if the Rescripts of 1839 and 1856 mark the ascendancy of Western influence in the Ottoman Empire, they also mark the rise to power of a westernizing party within the Ottoman official class. Without the activity of the Westernizers, who became prominent not only in the Ottoman Empire, but also in Persia, Tunis, and elsewhere, it is difficult to conceive of these Middle Eastern polities adopting and persevering in such a line of policy. These men wholeheartedly believed that the salvation and prosperity of their societies—as well as their own security from the murderous whims of despotic rulers—lay in introducing and seriously applying the European ideal of a *Rechtstaat*, of a government of laws not men. This became the goal which political leaders desired, or professed to desire, during a period which extends from the proclamation of the Rose Chamber Rescript to the mid-thirties of the twentieth century, when the triumph for the time being of Nazism, and the later rise of the Soviet Union to superpower status, made other political ideals more alluring.

The state of mind of the reformers, their ideals, and beliefs are well illustrated in a book written in 1867 by a Tunisian figure, Khayr al-Din al-Tunisi (d. 1890). Khayr al-Din himself was a prominent reformer in Tunis, but the reforms which he helped to introduce were cancelled after a few years, following a popular revolt provoked by the reformed institutions, the more oppressive bureaucracy which they entailed (as described by the French observer quoted earlier), and the more onerous taxes associated with them. The introduction to this book, which was also published in a French translation under the title *Réformes nécessaires aux états musulmans*, is an apologia for the policy of reform, and a plea for its absolute necessity if Muslim states are to survive, let alone flourish. Khayr al-Din appeals to Aristotle's authority in arguing that it is a grave mistake to substitute for the law one individual acting according to his will. Khayr al-Din is eloquent on the blessings which attend the rule of law. Where it obtains, subjects repay this blessing by engaging in all the activities which are the foundation of material well-being:

The craftsman, for example, must feel secure against being despoiled of any of the fruits of his labour or hampered in certain aspects of his work. What does it profit a people to have fertile lands with bountiful crops if the sower cannot realize the harvest of what he has planted? Who then will venture to sow it? Because of the faint hope of the people in many lands of Asia and Africa you find the most fertile fields uncultivated and neglected. There can be no doubt that the hostile action against property cuts off hopes, and with the severance of hope comes the severance of activities until finally destitution becomes so pervasive that it leads to annihilation.[3]

Another, certainly more prominent, reformer at the centre of the Empire, Midhat Pasha, expressed himself even more emphatically. In a draft rescript to introduce the Ottoman Constitution of 1876 (which Sultan Abd al-Hamid in the event rejected), Midhat wanted unambiguously to link the welfare of the Empire with its adoption of Western institutions: 'The Ottoman state', he wrote, 'being part of the European community, it has to follow the same methods that they use in order to be on an equal footing with its neighbours in the way of progress.' European civilization and its political ways became the ideal, proclaimed *de rigueur* by Middle Eastern rulers even when they had not the least intention of living up to it. Thus, Ismail, the rapacious and despotic ruler of Egypt, at the inauguration of the Suez Canal in 1869 proclaimed his country to be henceforth a part of Europe. For a brief moment after the First World War, when Western power and influence seemed unchallenged in the Middle East, this ideal dominated political discourse and rhetoric among the Westernized and Westernizing rulers.

Both Rescripts, that of 1839 and that of 1856, endeavoured to give solemn expression to these European ideals in an official document emanating from the highest authority—an authority whose power had been unbounded, and which now seemed to be committing itself to the limitation of this very power, and promising to put itself under the restraint of the laws. The Rose Chamber Rescript promised legislation which would guarantee 'perfect security for life, honour, and property', and which would set up a regular system for the assessment of taxes, as well as a regular system of conscription. 'If there is an absence of security for property', the Rescript declared, 'everyone remains indifferent to his state and his community', while security increases devotion and love for both. Muslims and non-Muslims equally, the Rescript promised, will be able to enjoy, and freely dispose of, their property; and they will, 'without exception, enjoy our imperial concessions'. The Rose Chamber Rescript thus not only promises the establishment of a *Rechtstaat*, it also gives a hint that under the new dispensation there would be equality of treatment between Muslim and non-

Muslim. Whatever the practical consequences or implications of this hint, it was bound to jar on traditional attitudes and expectations long unquestioned and taken for granted. That the ruler should bind himself to bow to the rule of law, thus implying and conceding that government had previously been unjust and oppressive, was disconcerting enough. Even more disconcerting was the use of language which put Muslim and non-Muslim, however circumspectly, on an equality. Such departures from customary attitudes was bound to shock, if only because the Rescript and its promises were not the outcome of any popular demand, but rather simply embodied the ideals of a narrow group of Westernized officials. Henceforth there would be a standing divergence between the universe of discourse familiar to the people, and that of their rulers, who became accustomed to speak an alien political idiom whose relevance to the actual conditions of the Middle East continued to be highly doubtful. One can go further and say that the formulation of political issues in this particular idiom gave position and power to the Westernizers at the expense of the hitherto absolute ruler, and enabled these Westernizers themselves to exercise a similarly absolute power over the population, in order to lead them into unfamiliar paths which they were reluctant or even unwilling to tread.

A case in point is the Imperial Rescript of 1856. This document considerably amplified and pointed up the hint in the earlier Rescript about equality between Muslim and non-Muslim. The new Rescript laid it down that 'Every distinction or designation tending to make any class whatever of the subjects of my empire inferior to another class on account of their religion, language, or race shall be forever effaced from administrative protocol.' The Rescript promised equality of admission to schools and public employment, as well as equality in military service. The equality of Muslims and non-Muslims came to be enshrined in the public law not only of the Ottoman Empire, but also of other Middle Eastern polities. It was a departure fraught with numerous grave and far-reaching consequences. These were Islamic polities whose cohesiveness had rested on Islamic solidarity. This solidarity in turn rested on the legal and political superiority of the Muslim over the non-Muslim, on the pride which Muslims felt in the state being peculiarly their own state. If legal and political equality were now instituted, would it be possible to find a substitute for the cohesiveness and solidarity necessary to hold the body politic together? Would a Muslim fight with the same fervour to uphold a state which was denying its traditional *raison d'être* and proclaiming its religious neutrality? In striving to be non-Islamic would the state not in effect become anti-Islamic? Some of the ramifications of this problem were seen in all their acuteness by Jevdet Pasha. The morale of the Ottoman army rested on the Islamic solidarity which animated the troops. Should the army conscript non-Muslims,

in time of need, how could the Colonel of a mixed battalion stir the zeal of his soldiers? In Europe indeed, patriotism has taken the place of religious devotion . . . their children hear the word 'fatherland' while they are still small, and so years later the call of patriotism has become effective with their soldiers. But among us, if we say the word 'fatherland' all that will come to the mind of the soldiers is their village squares. If we were to adopt the word 'fatherland' now, and if, in the course of time, it were to establish itself in men's minds and acquire the power that it has in Europe, even then it would not be as potent as religious zeal, nor could it take its place. Even that would take a long time, and in the meantime, our armies would be left without spirit.[4]

If the state now professed to be indifferently the state of Muslim, Christian, and Jew, what then was the principle which defined it, and determined its boundary? Hitherto the Muslim state was *dar al-islam* pure and simple, that is wherever Muslim power had sway. If this was no longer the case, territoriality as determined by, say, international treaties would have to define the state and describe the reach of its sovereignty, and to territoriality would have to be joined the principle of nationality, such that the inhabitants of a territory where a particular state was sovereign would be generally deemed by virtue of residence to be the nationals of such a state. The adoption of this modern European principle of international organization implicitly followed from the admission of the Ottoman Empire to the concert of Europe, and from the subsequent membership of other Middle Eastern states in a society of states whose organization and intercourse was regulated by international law, which Western legal traditions had fashioned in response to specifically Western conditions. But Western conditions were sharply different from Middle Eastern ones. In the West, there was a multiplicity of polities of long standing each with its own particular history and highly aware of its own specific institutions. In these conditions, the idea of states constituting a kind of society is accepted as naturally as are territorial boundaries, and allegiance based on residence. In the Middle East, on the other hand, these are felt to be tinged with artificiality. Turkey was established as a republic with brand-new boundaries following the destruction of the Ottoman Empire. It was to be the country of the Turks, supposed to be linked by a common past, a common language, and a common culture. But it is still very doubtful whether a Turk who is not a Sunni Muslim is considered by his fellow nationals, or considers himself, a Turk. Egypt, in contrast to the Turkish Republic, has been for centuries a reasonably well-defined geographical entity. But as a modern state, it is the creation of Muhammad Ali, who was formally an Ottoman subject as were also the inhabitants of the territory under his rule. It was only after the First World War that an Egyptian nationality came to be defined and recognized. Given the relative territorial stability of Egypt, the character of the Egyptian state and of Egyptian nationality ought to have been fairly clear. This, however, is not the case. Egypt contains a substantial Christian element, the Copts, whose status as Egyptians in the body politic is equivocal and uneasy. Again, however well-defined Egypt seems to be as a state, its self-identity has fluctuated. Ismail, as has been seen, considered in 1869 that its destiny lay henceforth with Europe. After the First World War, Egyptian political leaders were emphatic that Egypt was an entity on its own, sharply distinguished from its neighbours as well as from Europe. During this period, however, the King of Egypt attempted to gain for himself the Caliphate of which the Turkish Republic had divested the Ottoman dynasty. After the Second World War, Egyptian leaders saw Egypt as an Arab country. Indeed, for a decade or so after 1958 the very name of Egypt disappeared from the political map, being replaced by the United Arab Republic.

The full consequences of the principles exemplified by the Imperial Rescript of 1856—principles that most Middle Eastern states tried, or merely professed, to follow—took a considerable time to become fully evident. But there was no mistaking the communal tension which became manifest in the Ottoman Empire following the promulgation of the Rescript. Belief in the policy of the Rescript was confined to a small group of public men at the centre of affairs, Westernized in outlook, very conscious of the acute predicament of the Ottoman state, and exposed to the pressures and persuasions exerted by European powers, both friendly and hostile. They looked upon the new course as a necessity. But the Muslims at large felt shock at such a humiliation: they were offended by what they saw as the insolent claim to non-Muslims, who had hitherto been under protection, to have a say, as of right, in the affairs of the Muslim state; and they felt threatened in their traditional claims and position. The non-Muslim communities—of whom the Christians in their various denominations formed the over-whelming majority—saw this promise of equality as the gift of their now powerful European co-religionists. As Jevdet Pasha pointed out, they could not be expected to feel for a Muslim state the loyalty which Muslim generations had felt and transmitted to their descendants over twelve centuries. The notion of their being equal citizens with their fellow Muslims in an Ottoman state, with everyone professing loyalty to a shadowy and abstract Ottomanism, left them, rightly sceptical. They were not over-eager to discharge the duties of citizenship within a state where so recently they had been subjects with an inferior status, nor would their Muslim fellow citizens have put much credence in their sincerity or willingness to do so. As for the promise of equal rights, this could not but raise the expectations of the non-Muslims, while it aroused the fears of the Muslims. This was, in short, a very uncomfortable and at times dangerous state of affairs. The fears and expectations aroused by the 1856 Rescript had a great deal to do with the so-called Kuleli conspiracy of September 1859 in which some army officers and *ulema* sought to assassinate the Sultan and overthrow the government on the ground that they were endangering the state by their pro-Western proclivities, their corruption, and their readiness to enhance the position of the Christians. The Rescript should also be kept in mind in any explanation of the massacres of 1860 in Damascus and the Levant. The successful Young Turk Revolution led to the publication, fifty-two years after the 1856 Rescript, of another Imperial Rescript, in August 1908, in which Sultan Abd al-Hamid was made to assert again that 'it is absolutely necessary that the rights of all our subjects be strictly guaranteed'.

Although natural and positive laws prescribe that in justice there is no difference whatever between one individual and another, or one class of the population and another, and that all must receive equal justice, it is true that for some time, and contrary to our wishes, the application of this prescription has on occasion been lax.

The Rescript then reiterated:

All our subjects, of whatever *millet* or religious confession, are equal in their rights and duties before the state and in the enjoyment of their individual liberty.

The advent of a Young Turk regime inspired and controlled by Western-educated officers who were widely suspected of irreligion and of wishing to subvert Islam by making non-Muslims the equals of Muslims, again raised communal tensions, as witnessed by an Armenian massacre in Adana, an anti-Jewish outbreak in Baghdad in the months immediately following the Revolution, and similar incidents elsewhere.

Whatever its unforeseeable advantages or drawbacks, equality exacted sooner or later an inevitable price from the non-Muslim denizens of Middle Eastern states. Legal and political equality—however formal and empty of substance—made it difficult, if not impossible, to justify the *millet* system, with its communal institutions which acted as a shield to protect its members from the demands, and the usually unwelcome attentions of the government. If equality was now the rule there was no need for special privileges and for institutions to secure these. Under the new dispensation *millets* became minorities, since equality necessarily meant that one citizen, whatever his religion, was to count no more than another citizen. This is indeed the logic of democracy, but the logic becomes somewhat oppressive when majorities and minorities are immutable. The majority is thus perpetually dominant, and imposes its norms and views on the minority, while the minority has little or no interest in maintaining the body politic of which it is ostensibly a member. A minority in such circumstances remains as much outside the *pays politique* in a modern *Rechtstaat* as the *millet* had been when the state followed the laws of Islam. The minority may have a few representatives in parliamentary assemblies. But such representatives cannot possibly sway, or prevail against, the overwhelming numbers who belong to the majority religion. Again, should they attempt to act as the representatives of a minority, their position becomes false and awkward, precisely because the polity is deemed to be nationally homogeneous, and to proscribe divisions based on religion. The recognized status of the *millet* is replaced by the insecurity and tension which the discrepancy between democratic rhetoric and political reality must generate. Nor is the disappearance of a recognized, albeit subordinate, status compensated by the occasional appointment to minor political office—for example, of what might be called the statutory Copt to head an Egyptian department of state.

For *millets* to become minorities was one possible outcome of Westernization. Another was for *millets* to become, or to look upon themselves as nations. The idea that humanity is naturally divided into nations, and that each nation is entitled to an independent national territory, is a Western notion which came to the Middle East with other Western innovations. It spread very early in the Balkans among the Christian subjects of the Ottoman sultans. Indeed, the rebellion of a *millet* in the name of the new ideal of nationality occurred a few years before Mahmud II embarked in earnest on military modernization. Greek-speaking areas and groups had come under Ottoman domination in the fifteenth century. They were looked upon, and they looked upon themselves, as the Rum *millet*, the Romaioi, i.e. those Orthodox subjects of the Roman emperor in Byzantium whose rule had been ended by the Ottoman conquest of Constantinople in 1453. In the decades immediately preceding the revolt of 1821, some Westernized members of this *millet* were influenced by European thought and by the Western admiration of ancient Greek art and literature. They came to look upon their community as the heirs of Themistocles, Pericles, Socrates, and Plato—figures from a past which Orthodox teaching had hitherto condemned as pagan. The Greek nation was to arise anew and resurrect the glory that was Greece. This radical and revolutionary change in self-view, transforming the Rum *millet* which had lived under Ottoman subjection since the fifteenth century, into the Greek nation entitled, *ipso facto*, to political independence, makes of the Greek-speaking subjects of the sultan the first Middle Eastern group, whose intellectual classes were profoundly affected by the ideology of nationalism, and who thus underwent a species of westernization.

If *millet*s were now to be seen as nations—according to the nationalist meaning of the term—then the tensions generated within the traditional Middle Eastern polity would be even greater than those which resulted from the attempt to establish equality between Muslims and non-Muslims, and the consequent transformation of *millets* into minorities. The nationalist claim to independence was a direct challenge to the existing regime which was bound to be forcibly resisted. The claim may be successfully enforced: as it was in the case of Greece, Serbia, Romania, and Bulgaria; or it may fail: as it did in the case of the Ottoman Armenians, and, so far, in that of the Kurds in Turkey, Iraq, and Iran. (The Kurds, though not a separate *millet* in the Ottoman sense of the term (since they are Muslim) have come, in the eyes of their political leaders at least, to be considered a nationality.) In any event, a peculiarly unpleasant kind of conflict resulted in which no dividing line was drawn between armies and civilians, and in which settled populations of all kinds became simultaneously or in turn victims of massacre, flight, and expulsion: ever since the Greek Revolt,

Greeks, Turks, Bulgarians, Armenians, Kurds, Arabs, Jews, Maronite Christians, Druse, and Shiites have all been offered up at one time or another as sacrifices to the ideal of national homogeneity. If the Imperial Rescript of 1856 was designed to diminish the attraction of nationalism to the *millet*s, then its failure quickly became manifest. The very transformation of *millets* into minorities which the Rescript signified itself made nationalism attractive and sometimes irresistible.

Decades, then, of intensive military, legal, and administrative reform in the Ottoman Empire, Egypt, Tunis, and (to a much lesser extent) Iran did not result in greater military security, better or more economical administration, stable or less precarious public finances, a more contented or less restless population. It was not the unreformed Ottoman Empire, but a polity which had already undergone a great deal of reform which the Tsar called the Sick Man of Europe. No amount of military modernization seemed sufficient to preserve Middle Eastern states from military defeat, or foreign encroachment: in 1913 a coalition of Balkan states which had seceded from the Ottoman Empire in the course of the nineteenth century succeeded in defeating the Ottoman Army and almost in occupying Constantinople. Tunis was occupied by the French in 1881, Egypt by the British in 1882, Tripoli by the Italians in 1910, while Persia became the subject of a virtual partition between Russia and Great Britain in 1907. Modernization by its very nature entailed greater and greater public expenditures in order to pay for a greatly enlarged bureaucracy, to import costly weapons, and to carry out various enterprises which governments now persuaded themselves they had to undertake. Domestic taxes were by no means sufficient to cover the enormously increased outlays, and recourse had to be had to loans, whose burdens became increasingly onerous: the Ottoman Empire in 1875 and Egypt in 1876 had, successively, to declare themselves bankrupt. Instead of bringing amelioration or alleviation, reform seems to have only intensified and complicated the crisis it had been meant to tackle. But the pattern established by the activities of the nineteenth-century reformers has proved durable. Indeed, subsequent developments have served prodigiously to increase the centralizing proclivities of Middle Eastern governments, and to instil in them the ambition to regulate an ever increasing range of social activities. Over-administration, which stifles energies and kills initiative, has resulted. Its most immediate visible sign is a proliferating, intricate, burdensome, meddling bureaucracy. Every new regime denounces it for corruption and vows to purge and reform it, but soon enough realizes it to be beyond reform, and wellnigh beyond control.

3

Constitutionalism and its Failure: I

SOME two years before the publication of the Rose Chamber Rescript of 1839, a prominent official, Pertev Pasha, who was Secretary for Foreign Affairs, displeased his master Mahmud II. He was dismissed and exiled. Then, very shortly afterwards, the Sultan ordered him to be executed, and this was done by the traditional method of the bowstring. Whether in the Ottoman Empire or elsewhere in the Middle East, it was indeed the case that 'the neck of a servant of the sultan is thinner than a hair's breadth'. But the 'security for life, honour, and property' which the 1839 reforms promised, changed the fortunes of office-holders in the Ottoman Empire, as modernization in due course did, for a time at any rate, in other Middle Eastern states. Officials were, in fact, the clearest beneficiaries of the modernization which they advocated and carried out. The incubus of sudden confiscation and execution was lifted from them. Indeed power flowed to them, their numbers increased, and with increasing numbers came increasing importance, and a decreasing possibility that either the ruler or the subject could exert control over the official apparatus. A European observer intimately acquainted with the upper reaches of the Ottoman bureaucracy draws this rather hostile portrait in a book published in 1868. The officials, he wrote,

form a powerful corporation which, possessing the advantages of a relatively superior erudition and versed in the routine of administration, has easily been able to usurp and conserve a preponderance, over the other bodies of the state. Its political attributions extend to all branches of administration and thanks to it, this corps has acquired a limitless power and influence . . . Not content with the legitimate exercise of their power in the ministries which constitute their domain and in the sphere of their attributions, these [clerks] have seized power by spreading their ramifications and meddling in the smallest detail of governmental organization.[1]

But this considerable improvement in the situation and prospects of the officials itself began to evoke discontent and disaffection among those who had not themselves directly experienced the full weight of the Sultan's despotic power and the fear which he inspired in his servants. What they saw around them was a complicated bureaucratic machine imitating European models—a machine which did not work very well, which certainly did not serve to raise the Empire to the level of European civilization, but which allowed a few ministers favoured by the Sultan to enjoy extensive powers which had previously been his alone to exercise. As one of these critics— himself an official—Namik Kemal, put it, Istanbul now had many sultans; only they did not bear the title. Criticism of the reformed institutions thus emanated from members of the same official class which had earlier provided the motive force for modernization. Indeed, the critics were themselves the products of modernization, officials who had been exposed to European ideas because the needs of the state now required that some at any rate of its servants should receive a European-style education. Opposition to the *tanzimat* regime was a dialectical outcome of the *tanzimat* themselves.

Of those who launched new currents of thought, Ibrahim Shinasi (1826–71) was a clerk in the Artillery, who managed to learn French from a French officer who had converted to Islam, and who then went to study in France on a scholarship. A protégé of Reshid Pasha, the Grand Vizier who was the great proponent of the *tanzimat*, he combined literature and newspaper editing with his official work. His official career had its ups and downs, and in 1863 he was dismissed from office, possibly for having discussed in his newspaper the principle of no taxation without representation. The following year, Shinasi fled to France, perhaps because he was involved in a plot against the Grand Vizier. He left the editorship of his newspaper to a younger man who was an admirer of his and who had the same career pattern as himself.

This was Namik Kemal (1840–88) who, like Shinasi, combined an official and a literary career. Descended from an aristocratic family, he was taught French at home, and at the age of seventeen entered the Translation Office of the Sublime Porte. He was attracted to Shinasi as a literary innovator, but as in so many other cases, desire for literary innovation went hand in hand with desire for change which would make imperial institutions approximate to the political ideals which Western literature projected. These views were articulated in Namik Kemal's contributions to Shinasi's newspaper before and after the older man's exile. His political attitudes also found vent in membership of a small secret group composed of like-minded young officials, almost all of whom had at one time or another worked in the Translation Office of the Porte. In the summer of 1865 they met at a picnic in the Forest of Belgrade near Istanbul and decided to form a Patriotic Alliance, modelled on the Carbonari, in order to save the country from arbitrariness at home and defeats abroad. The policies which they so much resented they associated with the Grand Vizier Âli Pasha, and his colleague Fuad Pasha (both of whom had themselves begun their official careers in the Translation Office), who, together with their patron Reshid Pasha, had been the chief promoters of the *tanzimat*. The Patriotic Alliance did not accomplish anything much, but its members may be seen as heralding the discontent of the new official and intellectual classes which the *tanzimat* had brought forth. Namik Kemal and his friends are also precursors in seeking to establish that a new beginning was absolutely necessary if prosperity at home and safety abroad were to be ensured. In the spring of 1867 following an article on the affairs of Crete—where a rebellion of Greeks against Ottoman rule was in train—and a report of criticisms of the Ottoman government made from his Parisian exile by an Ottoman grandee, Namik Kemal's paper was closed, and he himself transferred to a provincial post. But Namik Kemal would not vegetate in a provincial exile, and soon managed to leave for Paris.

Another official who also felt the displeasure of the authorities in 1867 was Abdul Hamid Ziya Bey (1825–80). Like

Shinasi, he was a protégé of Reshid Pasha, who had procured for himself a post in the imperial palace. He learned French and began to publish Turkish translations of French books. Owing to the disfavour of Âli Pasha and Fuad Pasha, his bureaucratic career was a chequered one, and in his discontent he became an associate of Namik Kemal and his circle. Like Namik Kemal he was banished in 1867 to a provincial post, and like him decided to leave for Paris.

What drew them to Paris was an invitation by that Ottoman grandee whose criticism of the Ottoman government had shortly before circulated in Istanbul. This was Prince Mustafa Fazil, a grandson of Muhammad Ali, Vali of Egypt, and brother of the then ruler of Egypt, Ismail Pasha. Mustafa Fazil had served in high office in Istanbul, had fallen out with Fuad Pasha, and left for Paris. Whether because of this quarrel or of his disappointed expectation to be recognized by the Sultan as his brother's heir, Mustafa Fazil became an ostensible opponent of the regime, proclaiming its faults and shortcomings, and suggesting what he considered to be necessary reforms. His criticisms and remedies were set out in an open letter to the Sultan—whether written by him or not is here immaterial—which appeared in Paris in French in March 1867, and which Namik Kemal and his friends shortly afterwards translated into Turkish, printed, and distributed in Istanbul.

The open letter asserted that in the Ottoman Empire there were only two classes, the oppressors and the oppressed. The oppressors are those 'subordinate tyrants' who have arrogated to themselves the Sultan's unlimited powers, and whose despotism brought to mind the condition of France before the Revolution of 1789. Mustafa Fazil here quotes what Louis XV is supposed to have been told by a political observer, namely that no one in his kingdom, however exalted his position, could consider himself immune from the vengefulness of a minister; and no one, however lowly his position, could escape the vengefulness of a petty tax-collector. Under such conditions the oppressed suffer from an 'incurable moral cowardice', and their minds become atrophied under a regime which leaves nothing to the citizens' initiative.

Industry, commerce, and agriculture, everything decays in consequence. This decline, Mustafa Fazil is careful to point out, is in no way the fault of the religion, but entirely that of the political system. To prosper, the citizens have education. But to build schools is not enough: liberty is the first and foremost schoolmaster of the people. This is the lesson of Europe in modern times, of France, of Italy, of Austria, and of Prussia. To bring home his point, Mustafa Fazil listed other parts which were then more or less tenuously part of the Ottoman Empire and which had now followed the logic of European progress and established constitutional and representative assemblies. These were Serbia, Moldavia and Wallachia, Egypt, and Tunis. His advice therefore was to constitute in each province a freely elected assembly which would inform the Sultan about the condition of his subjects and thus enable him to realize his benevolent intentions. Delegates appointed by these assemblies should be summoned periodically by the Sultan to Istanbul where they would explain the condition of the people and give expression to their wishes.

The themes of this pamphlet were adopted by the Young Ottomans, as the critics of the regime who congregated in Paris under Prince Mustafa Fazil's patronage called themselves, in imitation of current European usage. These themes were repeated and developed in Turkish periodicals which the *émigrés*, financially helped by Mustafa Fazil, and by his brother and rival Ismail, in furtherance of their respective political ambitions, published for some two years in various European cities.

Thus Namik Kemal came to argue that constitutional and representative government, as well as the division of powers which characterized the most progressive European regimes, could be introduced into the Ottoman Empire because in reality these were original features of Ottoman, and more generally Islamic, government. In an article of 1868 Namik Kemal declared that before the Ottoman system was subverted by the recent centralization, it had been characterized by the division of powers. The Sultan and his viziers were the executive, the *ulema* were the legislative, and the Janizary corps represented the people and acted as a check on the executive.

Another Young Ottoman, Ali Suavi (1838–78), was more emphatic in deriving popular sovereignty from Islamic tradition. He referred to incidents in early Islamic history in order to prove that the Prophet and his immediate successors practised consultation in public affairs, and that parliamentary government was thus Islamic in origin. The partisans of constitutionalism quoted in particular two Koranic verses which, it was argued, enjoined consultation on the believers, even though neither was really germane to political issues. The first, 'and consult with them upon the matter' (verse 159, sura III), expresses God's approval of the Prophet's leniency towards those who had deserted in a battle against his Qurayshite enemies; while the second, 'and consult together in kindness' (verse 6, sura LXV), relates to the proper treatment of divorced women, and enjoins upon the husband to consult with the divorced wife about the suckling of their child.

Suavi went much further than his fellow Young Ottomans in appealing to the Koran and to Islamic sources in order to argue in favour of popular sovereignty and of the right of popular resistance. Thus he made use of the Koranic injunction enjoining the good and forbidding evil to argue that it was a duty incumbent particularly on the *ulema* to redress wrong and resist oppression within the Muslim community. He cited the example of the second Caliph, 'Umar, who asked the community to correct him should he be in error, and the determination of two of the Prophet's Companions to do so, if necessary by the sword. And Suavi went on, in an article published in one of the newspapers which the Young Ottomans were bringing out in exile:

O ye who desire justice! If you want to go about nodding your heads like snails, tyrants will never allow you to raise your voice. You are slaves.

If, on the other hand, you take to the sword and show your presence in the field of honour you will stand up against tyrants: you are human beings, you are free.

O people. How long are you still going to believe that a *Mehdi* shall appear and save you?

Do you think that the enemies who are in charge and who are free of question and responsibility will abandon what profits they draw out of you and begin to favour you?[2]

Suavi, who also did not flinch from advocating assassination in a righteous cause, seems to have been fully in earnest about his doctrine. Having returned to Istanbul in 1876, he soon came to consider the new Sultan, Abd al-Hamid, unworthy of his position. He organized a conspiracy to depose Abd al-Hamid and reinstate as Sultan his elder brother and immediate predecessor, Murad. With this in mind, on 20 August 1878 he led a crowd into the grounds of a palace where Murad was kept and attempted to release him. The police were called, and in the ensuing mêlée Suavi was killed, together with twenty of his followers. If Suavi was sincere in finding in Islam a warrant for radical political action, other reformers, aware of the deep attachment of ordinary Muslims to their religion, simply used Islam as a means of propagating political notions which they favoured, but which they knew were really alien to Islam. Thus one of the leaders of the Patriotic Alliance, Mehmed Bey (1843–74), graduate of the Ottoman School in Paris, member of the Translation Office, the son of a Vizier, nephew of a Grand Vizier and son-in-law of another, imbued with radical and revolutionary ideas, would put on once a week the dress of a man of religion in order to visit mosques and advance the cause of reform. Another reformer, the Persian Malkam Khan (1839–1908), a devotee of the religion of humanity, clearly saw that reformist ideas, whether in religion or politics, could make no headway in a Muslim country if presented in a Western garb. As he revealingly wrote in an English periodical, Islam is an ocean in which may be found support and confirmation for any innovation which it is desired to advocate.

As to the principles which are found in Europe, which constitute the root of your civilization, we must get hold of them somehow, no doubt; but instead of taking them from London or Paris, instead of saying this comes from such an ambassador, or that is advised by such a government (which will never be accepted), it will be very easy to take the same principle and to say that it comes from *Islam*, and that this can be soon proved. We found that ideas which were by no means accepted when coming from your agents in Europe, were accepted at once with the greatest delight when it was proved that they were latent in Islam.[3]

Those who deliberately made use of Islam, like Mehmed Bey or Malkam Khan, in order to further the cause of constitutional and representative government, were most probably a minority among the educated and official classes. Far more numerous were either those who sincerely believed that true Islam was compatible with, or actually required, popular sovereignty and limited, constitutional government; or those who were convinced that progress simply required this kind of government, were indifferent whether or not it was compatible with Islam, or even prepared for its sake to abandon Islam as the state religion.

Thus a prominent and influential figure among the Young Turks—as the opponents of Abd al-Hamid chose to be called—Murad Bey al-Daghistani (1853–1912), summing up the Young Turk creed in a book published in 1897, declares that Islam was not the direct cause of the weakness of the Ottoman Empire, and does not constitute an obstacle to its recovery. Such recovery depends on the establishment of a constitutional regime for which the Empire was ripe, just as the present decay and corruption is solely attributable to the nefarious effects of Sultan Abd al-Hamid's despotism. A decade and a half earlier, about 1880, amidst the effervescence occasioned by the deposition of the Khedive Ismail and the beginnings of the intervention in Egyptian politics of Colonel ʿUrabi and his fellow-officers, Saʿd Zaghlul (1857–1927), who became prominent in Egyptian politics during the first three decades of the twentieth century, wrote an article on the relation between constitutionalism and Islam in *The Egyptian Gazette*, of which he was the literary editor. Divine law, the radical young Zaghlul wrote, does not allow tyranny. It requires that rule should be limited by tradition and law, and that there should be men ready to set right the ruler should he deviate from the true path:

It is for this reason that our Lord 'Umar [the second of the well-guided Caliphs] . . . asked the people, in his well-known address, to set him right whenever he erred in applying the rules of the [divine law], and for this reason God, the most high, said: 'Let there be formed among you an *umma*—[i.e. a group]—who call for good deeds . . . and who prohibit evil'.[4]

This divine injunction, so Zaghlul argued in a fashion similar to Suavi's, meant that a group drawn from the Muslim community should arise with the duty to call for good actions and to prohibit evil ones, so that those who are tempted to transgress—i.e. the tyrants—should be hindered from evil actions.

That Islam was originally democratic and Muslim rule constitutional was also argued some two decades later by another Arabic writer, Abd al-Rahman al-Kawakibi (1849–1902), whose writings became widely read and influential. With the passage of the years, these original features of Islam, Kawakibi believed, were replaced by tyranny which venal and corrupt divines defended and abetted. Today, fear of the tyrant is the outcome of ignorance. If the people are educated, they will become aware of the fragile basis on which tyranny rests, and will thus be able to overthrow it and re-establish constitutional government, one in which every member of the body politic will have his say on public affairs.

Kawakibi's influence extended beyond the Arabic-speaking world. A Persian divine, Mirza Muhammad Husayn Gharavi Na'ini (1860–1936) adapted the arguments of Kawakibi's *Characteristics of Tyranny* (one of the two books which Kawakibi wrote, first published as newspaper articles in 1900) to a defence of constitutionalism and of the Persian revolution of 1906. The adaptation was necessary since the political theory of Shiism was centred on the Hidden Imam, the repository of all legitimate authority. In his absence tyrannical rulers are mere usurpers. The effects of usurpation can be removed, however, if rulers gain the approval of the divines who are the agents of the Imam, rule according to a constitution, and are responsible to a consultative assembly, which in turn is responsible to the people. Like Kawakibi, Na'ini attacks divines who have sided with, and defended, political tyranny in order to safeguard their own religious tyranny. And like Kawakibi he believes that the spread of education will end their power:

Unfortunately this is no solution to the problem of religious tyranny, and the evil *ulama* cannot be led to the truth. They are not going to terminate their stubbornness and their tyrannical attitudes. On the other hand, weak-minded common people are not able to differentiate between truthful religious leaders and these religious thieves . . . There is only one way to end this religious tyranny: political tyranny cannot be religiously legitimized for ever, and the co-operation of the religious authority cannot always be invoked as 'the protection of religion'. Therefore, at this end of the century when people are becoming awakened, there is some hope that the people's own awareness will eliminate the influence of religious tyranny.[5]

As may be appreciated, however sincerely meant, the belief in the compatibility, or even the confluence, of Islam and constitutionalism had no basis in traditional Muslim political thought or in the historical experience of Muslim states. Nevertheless, the argument became very influential among those who were in the vanguard of the constitutionalist movement, and eventually found its way into official documents. Thus, in the Imperial Rescript which promulgated the Ottoman Constitution of 1876, Sultan Abd al-Hamid II declared that this

fundamental charter establishes the prerogatives of the Sovereign, freedom, the civil and political equality of the Ottomans before the law, the powers and responsibilities of ministers and officials; the right of control exercised by Parliament; the complete independence of the courts; the effective balancing of the budget; and administrative decentralization in the provinces, while safeguarding the central government's functions and power of decision.

All these features, the Sultan went on to declare, were in agreement with the requirements of the *shari'a*.

Such a view, however, began to seem unsatisfactory, indeed untenable, to some of the Young Turks, the heirs of the Young Ottomans who became active during Abd al-Hamid's long reign (1876–1909). These Young Turks came to believe that Islam was reactionary, the incubus which inhibited progress, and that

Muslim divines were hand-in-glove with the Sultan's despotism and simply used religion as one of its bulwarks. As much for the sake of Islam as for that of the Sultan's subjects, Islam and the state had to be freed from each other. Ceasing to be a political instrument, Islam would be purified, and made into a religion which the individual freely accepted—or rejected—exclusively at the dictate of his own conscience. Simultaneously, state and society would be free to follow the path of civilization—which was European civilization.

This was emphatically the view of Mustafa Kemal, later known as Kemal Atatürk (1881–1938). The Greek occupation of Smyrna at the end of the First World War, the utter powerlessness of the Sultan and his government in Istanbul, controlled as they were by the victorious Entente Powers, persuaded this Ottoman general who had already shown his quality during the War, to organize armed resistance to this invasion of the fatherland. By dint of his prowess and his extraordinary powers of organization and leadership, he found himself within a short space of time the head of the effective government of Anatolia. This government exercised its power not from Istanbul, but from Ankara. There, a Grand National Assembly enacted in January 1921 a short Constitutional Law. The first article of this law declared that 'Sovereignty belongs unconditionally to the nation', and the second decreed that 'Executive power and legislative authority are manifested and concentrated in the Grand National Assembly and its government bears the name of the "Government of the Grand National Assembly"'. This was entirely incompatible with Article 3 of the 1876 Constitution which declared that Ottoman sovereignty 'belongs' to the eldest of the princes of the Ottoman dynasty. And in fact, a Sultan was still on the throne in his capital of Istanbul. This situation could obviously not last very long, and by 1924 the Ottoman dynasty had been stripped of all its prerogatives and banished from the territory of what a new Constitution now called the Turkish Republic. This Constitution, enacted in April 1924, followed the Constitution of 1876 in declaring Islam to be the religion of the state. But Mustafa Kemal—now President of the Republic—was determined to remove the slightest vestige of a link between state and religion. At its 1927 congress, the Republican People's Party, which he founded and headed, adopted a set of general principles, one of which stated that

The Party, by separating religious and philosophical beliefs from politics and the diverse complications of politics, bases its investigations, in the making of all the political, social, economic laws, in the establishment of all organizations, and in the meeting of all needs, on the principles that contemporary civilization accepts; in other words the Party considers the complete separation of religion from national and state affairs to be among its primary aims.[6]

Henceforth, secularism was to remain one of the principles of Kemalism, the official ideology of the Turkish Republic. An amendment of Article 2 of the Constitution in 1928 removed the mention of Islam as the religion of the state, and another amendment in 1937 declared that the Turkish state was a secular state. The same principle is affirmed in the Constitution of 1961 and in its 1982 successor.

The affirmation of the secularity of the state now firmly enshrined in the Turkish Constitution is one possible *terminus ad quem*—and certainly the most logical one—of the issue which arose in the second half of the nineteenth century regarding the relation between Islam, constitutionalism, and represenative government. But it has not been possible for other Middle Eastern states to adopt such a solution. All of them find it necessary to affirm that Islam is the religion of the state, or that the head of the state has to be a Muslim. Secularity apart, however, there remains the much larger question of the fortunes of constitutional and representative government in the Middle East—whether or not it was believed to be mutually compatible, or at least reconcilable, with Islam. Here, the experience of the Ottoman Empire and the Turkish Republic must loom large, but as has been seen from references made in Mustafa Fazil's open letter, representative institutions had in fact come into being earlier elsewhere in the Middle East. In extolling the benefits of these institutions, the Young Ottomans spoke of Tunis and Egypt—almost completely autonomous Ottoman provinces. The Bey of Tunis, Muhammad as-Sadiq (1859–82) granted a constitution in 1861. This grant followed attempts, carried out by a predecessor, the Bey Ahmad, to imitate Muhammad Ali of Egypt and acquire a modern army and navy. The attempt was a dismal failure. Its only result was that Tunis was caught in a web of international indebtedness. This meant that the influence of European Powers, and particularly France and Great Britain, grew and became wellnigh irresistible. In 1857, the consuls of these two Great Powers advised Ahmad's successor, Muhammad Bey (1855–9), to emulate the Sultan his suzerain and initiate reforms similar to those which had been edicted in Istanbul in the Imperial Rescript of the previous year. Muhammad as-Sadiq went even further and issued the Constitution of 1861. This Constitution set up a Grand Council partly appointed by the Bey and partly co-opted. It laid down that the Bey and his ministers were responsible to the Grand Council whose agreement was indispensable for the enactment of new laws, and the modification of old ones, as well as for their interpretation, for increases in expenditure, or for enlarging the army and the navy. This Grand Council was in the event packed with the followers and clients of the Bey's all-powerful minister, the Mameluke Mustafa al-Khaznadar who could then shelter behind this Council with all its extensive functions, and exercise without challenge an unrestricted and absolute power. The consequences were described by the same acute observer who has already been quoted in the preceding Chapter.

This supreme Council combines the powers of the Legislative Assembly, of the Senate, of the *conseil d'état*, of the court of appeal, and of the *cour des comptes*. The guarantees which the governed in constitutional states find in the division of power exist no more here, and one is at a loss to explain how the European consuls who . . . took part in drafting the Tunisian Constitution could have been accomplices in the monstrous creation of this supreme Council where all powers are mixed up . . . It is true that the life of the Beys is no longer threatened daily, but only on condition that their power and their intervention in public affairs become non-existent. At the side of the Head of State, reduced to the role of a do-nothing king (*roi fainéant*), a permanent committee composed of ten members of the supreme Council watches daily. This committee, called the committee of ordinary affairs . . . is made up of the most influential Mamelukes.

This is the crown of the whole edifice. The Tunisian constitution, the liberalism of which is lauded by some journalists, leads to the formation of a Council of Ten, a kind of vigilance commission similar to that invented by the French Assembly of 1849. Power, influence, wealth, all is concentrated in it, and the situation of [Tunis] is perfectly summed up in these lucid words which General Bonaparte addressed to the Egyptian population: Is there a beautiful woman? She is for the Mamelukes. A fine horse? It is for the Mamelukes. Money, fruitful trees, fertile land? They are all for the Mamelukes.[7]

This new regime, constitutional but not representative, signified no more than that absolute power, hitherto exercised by the Bey, was now transferred to his officials. The regime did not last very long. At the end of 1863, the rate of the *mejba*, a poll tax which was anyway unpopular, was doubled. The population of Tunis at that time comprised a large element of nomads and semi-nomads not easily controllable by a government which, however tyrannical its instincts, still lacked the necessary means for efficient and speedy suppression. A rebellion broke out, its battle-cry being 'No more Mamelukes, no more *mejba*, no more constitution'. The rebellion was eventually put down, but the constitution had to be suspended, and the suspension proved permanent.

Ismail Pasha, Muhammad Ali's grandson, who ruled Egypt from 1863 to 1879, did not have to contend with an entrenched official class which might aspire, as in Tunis, to transfer the substance of power from the ruler to itself. The ruler himself was the exclusive fountain of honour and power, the officials were his servants, to be raised, dismissed, exiled, dispossessed, or killed at his will. Nor did European Powers press him to grant an assembly. If, in 1866, Ismail did so, this was only an expression of his desire to give Egypt a modern, Western appearance, in the hope that this would increase his stature in the world, strengthen his position *vis-à-vis* his suzerain, the Ottoman Sultan, and perhaps, in certain circumstances, give a colourable aspect of legitimacy and popular support to his financial policies. In a rescript of 22 October 1866, he declared: 'Whereas the great uses and benefits of consultative assemblies have been observed in civilized states, I have been in hopes of constituting an assembly in Egypt, the members of which would be elected from amongst the people.' The assembly so established consisted of seventy-five members elected for three years. The electors were the *omdas*, the village headmen, whom the rescript described as 'notables appointed according to the wishes of the population'. It was, of course, notorious that these headmen were the creatures of the government. This assembly, wholly under the ruler's control, proved to be of no consequence. But during the last year of his reign, in 1879, Ismail contrived to bring it into the public eye. He was now burdened with large debts incurred through ambitious and grandiose policies. The European Powers pressed for the Egyptian public finances to be taken in hand by European controllers who would seek to reorganize them and arrange for the payment of the debts in a methodical and orderly manner. This meant a drastic curtailment in Ismail's income and in his prerogatives. To prevent this, Ismail convoked the assembly and had it protest against the European-sponsored schemes as being contrary to the interests of the Egyptian people. The quarrel, he calculated, from being one between him and his creditors, would become one between an oppressed Egypt and grasping, foreign, money-lenders. In his *Modern Egypt*,[8] Lord Cromer, who knew at first hand Egyptian conditions during the last years of Ismail's rule, has a chapter which describes with biting irony Ismail's clumsy experiment to 'call free institutions temporarily into existence as an instrument through whose agency he might regain his personal power, which was threatened by foreign interference'. 'It was', Cromer wrote, 'a curious sight to see Ismail Pasha who was the living embodiment of despotic government in its most extreme form, posing as an ultra-constitutional ruler who could not conscientiously place himself in opposition to the national will.' Cromer imagines Ismail reasoning to himself that since Europeans lay much stress on the will of the people and have large talking assemblies, termed Parliaments, they can be hoist with their own petard. Thus if a scheme were devised which would present matters to the British Government and public in a form to which they were accustomed, if their most cherished institutions were apparently copied in Egypt, if the Egyptian people were to express their own views through their own representatives, then the bait would take. An Egyptian Parliament should, therefore, be assembled. The representatives of the Egyptian people should express their devotion to the Khedive, and their satisfaction with his system of government . . . The members of the Egyptian Parliament must be left to devise their own [financial] scheme. That was essential. Otherwise, constitutional government would be a mere farce.

Ismail's stratagem, however, did not take in his creditors or the Powers who were now deeply involved in the Egyptian problem. Ismail was deposed by the Sultan, he went into exile, and his son Tawfiq (1879–92) succeeded him. A new phenomenon now appeared in Egypt, namely military intervention in politics. A few months before his fall, Ismail had also instigated some army officers to mutiny against a ministry to which he was hostile. The mutiny was successful and Ismail was able to dismiss ministers who were intent on securing financial solvency for the country by putting restraints on its ruler's powers. Tawfiq's accession seemed to open up opportunities for the acquisition by some of Ismail's erstwhile servants of some of the power which the departed ruler had for so long tenaciously held in his grasp. The officers whom Ismail had taught to mutiny, mutinied twice more against his son, in February and September 1881. On both occasions Tawfiq gave way, and it became obvious that ultimate power now lay with Colonel Ahmad 'Urabi and the other officers who led the mutiny. These officers, who were native Egyptians, had grievances against the Turks and Circassians who had the lion's share in military appointments and promotions. They were also, after the first mutiny, afraid for their life and security. But the success of their rebellion tempted them to embrace a much wider cause: namely the deliverance of Egypt from foreign interference and exploitation, whether by Turco-Circassians or Europeans, and the limitation of the absolute power of the ruler.

The first goal was undoubtedly popular in the country at

large. As for the second, it was shared by notables and officials who hoped that the power of which Tawfiq was divested would devolve on the assembly and council of ministers. In the months following the September 1881 mutiny, an alliance between mutinous officers, notables, and officials came into being, and it seemed as though Egypt would be governed by a responsible and representative government. In the event such hopes could not be put to the test. Throughout the first half of 1882 political tension was high in Egypt. Britain and particularly France were exerting a great deal of pressure on the new regime; the Sultan was also intervening in Egyptian affairs; and 'Urabi, who had become Minister of War and the moving spirit in the Government, was suspicious and jittery. But it was clear that he had become the supreme arbiter in Egyptian politics. Towards the end of May Britain and France required that he should temporarily retire from Egypt. In protest, the Ministry resigned and Tawfiq accepted its resignation. But 'Urabi's fellow-officers demanded his reinstatement, and a leader of the notables publicly informed Tawfiq that unless he agreed to reinstate the Minister of War his life was not safe. The upshot of the crisis was the bombardment by the British Fleet of the forts at the entrance of Alexandria harbour, following 'Urabi's defiance of repeated ultimatums to the effect that shore defences should not be fortified, and gun emplacements dismantled. 'Urabi declared war on Britain, and was himself proclaimed a rebel by Tawfiq. British forces landed in Egypt, defeated 'Urabi's army and entered Cairo on 14 September. For many decades to come, Britain was to remain the paramount power in Egypt. 'Urabi was tried, condemned, and sentenced to exile. It is of course impossible to say what the fate of Egyptian constitutionalism would have been had the British not occupied Egypt. The fact remains, however, that in the few months between the September 1881 mutiny and the bombardment of Alexandria in the following July, 'Urabi and his soldiers were the only effective power in Egypt, and that just as Tawfiq could not withstand their demands, so likewise no assembly could have done so. Again, the fortunes of constitutionalism in Egypt during the twentieth century give no warrant to the supposition that it would have prospered in the nineteenth century had there been no foreign intervention.

In the latter half of the 1870s the issue of constitutionalism and representative government came to a head not only in Cairo, but in Istanbul as well. At the centre of the Empire this too was occasioned by a crisis. The Ottoman Empire was unable to meet its financial obligations, and rebellion in the Balkans threatened to trigger intervention by Russia and other European Great Powers. Just as in the 1860s foreign threats to the Empire had led the Young Ottomans to produce their critique of Ottoman government and a remedy for its ills, so now some of the official classes once again began to advocate limitation of the Sultan's absolutism and the establishment of a parliament as the only way out of the crisis. The best-known, and indeed most important, of these was Midhat Pasha (1822–83), who had made a reputation for himself as an active and reforming provincial governor, and who had very briefly served as Grand Vizier in 1872. In August 1875 he was appointed Minister of Justice and made himself the advocate of constitutional reform in the Council of Ministers. He was not successful, and asked to be relieved of his post. In March 1876, an anonymous 'Manifesto of Muslim Patriots' was sent to various European personalities. It reflected the ideas of Midhat who was described in it as 'the enlightened and courageous head of the energetic and moderate party', and it was most probably drafted by him.

If instead of being subject to the extravagant caprices of a sovereign like Sultan Abdul Aziz, the population of Turkey were ruled by a good government, there would no longer be any question of quarrels concerning race and religion. If, instead of being governed by a despot who seriously believes himself to be the representative of God, inspired by Him, we were ruled by a wise Monarch, supported by a consultative Chamber, composed of representatives of every race and religion, Turkey would, without difficulty take the place to which the fertility of her territories and the intelligence of her people permit her to aspire.[9]

What, 'in a word' the 'Muslim Patriots' required was 'a Parliament on the English model'.

The dissemination of this manifesto was not the only sign that Midhat and his helpers were able to exploit new techniques of political action. There is evidence to indicate that Midhat incited, possibly by distributing money among them, *ulema* and students in religious schools in Istanbul—who numbered between fifty and sixty thousand—to demonstrate against the government for allowing Islam to be humiliated and fellow-Muslims massacred in the Balkans. Strikes and demonstrations by these students at a time when the Russian military threat to the regime was manifest induced the Sultan to dismiss his ministers and to appoint a new Grand Vizier. Midhat was included in the new administration as a Minister without portfolio.

The new ministers assumed office on 12 May. On 30 May, a military *coup d'état* engineered by the Minister of War with the active co-operation of Midhat and of the director of the military academy, and with the help or acquiescence of the Grand Vizier, the minister of marine, *sheikh al-islam* (the highest religious dignitary in the Empire), and other high-ranking officers and officials, deposed Sultan Abd al-Aziz and replaced him by his nephew Murad. As the sequel was to show, by no means all of these personalities were in favour of constitutionalism and representative government. But the ideas of Midhat and his supporters seemed now to be in the ascendant. The official announcement of Murad's accession declared that he had become Sultan 'by the grace of God and the will of the people', and Murad himself in his accession rescript affirmed that he was ruling 'by the favour of the Almighty and the will of my subjects'. Furthermore, Ziya Pasha and Namik Kemal, who supported Midhat, were now appointed secretaries to the Sultan. Three months later, however, Midhat and his colleagues decided that Murad, suspected of insanity, had to be deposed. To replace him they chose his brother Abd al-Hamid, after Midhat had obtained assurances from him that he would promulgate a constitution and would act as a constitutional monarch.

The year of the Three Sultans—as 1876 is known in Ottoman history—did indeed see the promulgation of a constitution, on

23 December. Its provisions were the outcome of a tussle for power between Midhat and the new Sultan. In this tussle Midhat was defeated. Had he won, the Grand Vizier would have been transformed into a Prime Minister presiding over a Council of Ministers who were chosen by himself, and who exercised collective responsibility. Such a provision would have meant a shift in power from the Sultan to the officials, in the same way as the Tunisian Constitution had earlier effected such a shift. Such a shift, however, would have meant that ministerial absolutism would simply have replaced the Sultan's absolutism. An example of what might have happened may be seen in the Muhyiddin plot. Ever since talk of a constitution became prevalent after Murad's accession, there was discontent in Istanbul, particularly among the *ulema* and the religious students, at the possibility that non-Muslims might sit together with Muslims in a representative assembly. At the beginning of August they sent a memorial to Midhat which declared:

We have subjugated the Christians and conquered the land with the sword, and we do not want to share the administration of the country with them or let them participate in the leadership of the government.[10]

The agitation continued into the autumn and Midhat ordered the police to investigate. They reported that Muhyiddin Effendi and other prominent *ulema* were its leaders, and were now under arrest. Midhat submitted the report to the Council of Ministers, and it was decided that they should be exiled without delay. Abd al-Hamid's subsequent treatment of Midhat himself mirrors this precedent exactly. The Sultan insisted that the Constitution should contain a clause empowering him to deport abroad anyone whom a police report suspected of constituting a danger to the safety of the state. Abd al-Hamid used this power to banish Midhat to Europe on 5 February 1877, less than two months after he had appointed him Grand Vizier.

Midhat's exile set the seal on his defeat by Abd al-Hamid. The new Sultan was obviously aware from the start that whatever a constitution and a parliament portended for the Empire, the real question was whether the final authority in the state would be his or Midhat's to exercise. Here was a situation somewhat similar to that of Tunis in 1861–83. In the earlier episode, the Constitution transferred the Bey's power to an oligarchy of officials. There is no reason to suppose that if Midhat had been successful in imposing his views, ultimate authority would not have likewise lain with the first minister and his associates. It is significant that during Midhat's brief tenure, the Sultan had reason to suspect that helped by his followers, particularly Ziya Pasha and Namik Kemal, the Grand Vizier was trying to establish an independent power-base for himself by recruiting a volunteer militia which would be a kind of private army under his control. In any case it is clear that the Parliament provided for in the 1876 Constitution would not have served as a check on executive power. The Parliament consisted of two chambers: a Senate and a Chamber of Deputies. The Senate was a body wholly appointed by the Sultan, while the Deputies were elected by the members of various provincial and local councils, who themselves were indirectly elected. These electors and their own electors in turn were far from thinking themselves the sovereign people. Their most vivid and most enduring political instinct was to obey the powers that be, embodied in the provincial governor and his subordinates. Deputies elected in such a manner were themselves on the whole unlikely to act as the people's tribunes. Rather, they would obey the fiat of the Sultan or of the Minister, whichever chanced to prevail.

This manner of electing the Deputies was said to be necessitated by the urgency of the case, and to be a strictly provisional arrangement. In the event, it was also adopted for the second session, which met at the end of December 1877 and was abruptly suspended by the Sultan on 14 February 1878. No more parliaments were elected until the Young Turk Revolution in 1908.

But even if elections had been direct, and even if, *per impossibile*, electors and representatives had been at home with this kind of politics, the Constitution of 1876, reflecting contemporary realities, gave preponderant power to the executive. Parliament could meet only when summoned by the Sultan, and could be suspended at his pleasure. Its members could neither initiate new legislation nor modify existing laws. They had to be content to vote on bills submitted on behalf of the Sultan. The only question at issue was whether the Sultan was to have the substance of power, or whether he was to be guided by his ministers. This was the object of the tussle between Midhat and Abd al-Hamid before the Constitution was promulgated. Midhat desired to make provision for a prime minister and a council of ministers collectively responsible for government. This was resisted by Abd al-Hamid, and instead of a prime minister, the Constitution provided for a Grand Vizier appointed by the Sultan and in effect responsible to him. It also provided for a number of ministers each appointed and dismissed at the Sultan's pleasure, and each responsible for his own department. The Sultan was declared to be the Sovereign of all Ottomans; he enjoyed wide civil, military, and administrative powers, and yet was also declared to bear no responsibility for his actions. The Sultan also had the unfettered power to call and suspend the Parliament, and appointments to the Senate were in his gift.

The Constitution attempted to deal with an issue which had come to the fore earlier, during the era of reforms—namely the position of non-Muslims in an Islamic state. The reformers sought equality in rights and duties between Muslims and non-Muslims. This, as has been seen, subverted the traditional state of affairs, but did not substitute for it a new stable relationship acceptable to both sides. The Young Ottomans who advocated constitutionalism blithely assumed that representative government would somehow dissolve the difficulty, but in a reply to Mustafa Fazil's open letter, also published in 1867, the Grand Vizier Âli Pasha put his finger on a fundamental difficulty which parliamentary government would face in a heterogeneous polity such as the Ottoman Empire:

The Ottoman Empire numbers twelve or fourteen nationalities, and unfortunately, as a result of the religious and racial hatreds which divide, above all, the Christian populations, none of these nationalities as yet shows great inclination to grant just and necessary concessions. If the representatives which they would nominate through elections were to be brought together today, such a national assembly would instantly give rise to all scandals imaginable.[11]

Though the Constitution now declared that Islam was the religion of the state, it also laid down that all subjects of the Empire were without exception called Ottomans, 'whatever religion they professed'. It also affirmed that all Ottomans were equal before the law, having the same rights and duties, 'regardless of religion'. When, however, elections came to be held, each religion was, by administrative fiat, given a ration of deputies. Though this meant that non-Muslims had a recognized position under the new dispensation, yet it also meant that this position was not essentially different from that which the *millet* system had secured. And this clearly went against the equality in rights and duties guaranteed to all Ottomans, irrespective of religion. In the upshot, practically no Middle Eastern state has managed effectively to marry constitutionalism with religious diversity.

The proclamation of constitutional and representative government in the Ottoman Empire in 1876 was, as has been seen, the outcome of a conspiracy in which the involvement of high-ranking officers was essential to success. Of these, Avni Pasha, the Minister of War, seems to have acted out of ambition; he was, if anything, opposed to making political innovations. But the Director of the Military Academy, Sulayman Pasha, a Westernized figure, was moved by the belief that Western political institutions were necessary for the salvation of the state. Sulayman Pasha and 'Urabi and his Egyptian fellow-officers may be considered as the precursors and harbingers of ideologically minded and conspiratorial officers who, in the decades to come, were to give Middle Eastern politics much of its characteristic flavour and style.

Abd al-Hamid is on record as explaining why he dismissed the Parliament in February 1878 and refused thereafter to convene a new one:

I made a mistake when I wished to imitate my father [Sultan Abd al-Majid], who sought to reform by persuasion and by liberal institutions. I shall follow in the footsteps of my grandfather [Sultan Mahmud].

Like him I now understand that it is only by force that one can move the people with whose protection God has entrusted me.[12]

It is doubtful whether the Sultan was right in saying that his father sought to reform by persuasion and liberal institutions, for, after all, Sultan Abd al-Majid's reign saw a vast centralization and extension of government, in which persuasion had no role to play. What is undoubtedly true is that, during his own long reign, Abd al-Hamid continued the policy of modernization and centralization which had been the hallmark of the *tanzimat* under the reign of his two predecessors, Abd al-Majid and Abd al-Aziz. But he was successful in reversing the tendency for power to flow from the Sultan to the officials—a tendency which, in a sense, had culminated in Midhat's *coup d'état*. Among his reforms, Abd al-Hamid expanded and improved military education. But the young officers graduating in increased numbers from the military academies necessarily became touched with the same Western ideas which had moved the Young Ottomans in their generation. They were rebellious at the concentration of power in one pair of hands, and felt the meticulous control which the Sultan attempted to exert over public affairs to be increasingly oppressive. They looked back with yearning and regret to Midhat's Constitution and Parliament. The very fact that they had been so short-lived served immeasurably to increase their attractions, and to persuade these young officers that all would be well if only the Constitution and the Parliament could be restored. Many of these young officers discussed and plotted in secret how to do this. When they were discovered, they were punished, most usually with exile to an outlying province such as Tripoli or Barbary, or else they fled abroad where they would publish anti-Hamidian newspapers or organize meetings at which would be discussed schemes to end despotism and establish a new regime where liberty, order, union, and progress would reign.

For about two decades, these groupings of Young Turks—as they came to be known to European publicists—persisted ineffectually in their attacks on the regime. Success came when a small number of young officers stationed in Salonika, the headquarters of the Third Army Corps, mutinied in July 1908. The mutiny spread to the Second Army Corps in Edirne. The Sultan sent a force from Izmir to put down the rebellion but these troops showed themselves to be in sympathy with the mutineers. Abd al-Hamid, who had been strangely passive in the face of the mutiny, now capitulated and agreed to the restoration of the 1876 Constitution, as the Salonika officers were demanding. The triumphant conspirators now proclaimed themselves as the Committee of Union and Progress. Very quickly, there was established a network of Committee branches in the provinces. The branch members belonged overwhelmingly to the official classes: young officers and bureaucrats who prided themselves on their enlightenment and saw in the revolution—for this is how the Salonika mutiny now came to be universally described—a great opportunity for them to rescue the Empire all at once from despotism, military weakness, and economic and social backwardness. A well-known Turkish journalist, Husayn Jahid Yalchin, who observed the Revolution, later on described in a remarkable passage the outlook of the Young Turks at the time when they so confidently took over the governance of an Empire teeming with fearful complications and faced with dangerous threats to its existence:

They were not very well educated . . . They had no idea of how complicated, enormous, and decisive were the dangers which awaited the Ottoman Empire . . . They only knew from their daily experience that Turkism and the fatherland were threatened with extinction, and the knowledge was unbearably painful. They believed that the only solution lay in terminating the despotism of the Sultan and in proclaiming liberty . . . They believed that in order to rescue the Ottoman Empire it would be sufficient merely to proclaim the constitution. They thought that all the complaints of the various elements proceeded from poor administration, from oppression, and from despotism . . . They believed that the constitution would solve everything. They believed that once the country experienced a change, thanks to the constitution, strong administrators and honest men would rise to power.[13]

The restoration of the 1876 Constitution was followed by elections at the end of 1908. Naturally, the new deputies were overwhelmingly in favour of the new order: the members of the Committee, working together with the provincial administrations, quick to respond to the wishes of the new masters, made sure that they would be. And as in 1876, the authorities assigned to the generally acquiescent non-Muslim communities what they considered to be their proper and due share of representation: only the Greeks raised objections. When the new Parliament met at the end of 1908, amendments were quickly carried through transferring the considerable powers which the Sultan had enjoyed under the 1876 Constitution to the Parliament. Abd al-Hamid did try to preserve some powers. He insisted that, as he was supreme commander of the armed forces, he should have the right to appoint the ministers of war and of the navy, but he failed to carry his point, and it was established that these particular ministers, like all the others, would be chosen by the Prime Minister, who, together with his colleagues, would collectively assume responsibility before the representatives of the people assembled in Parliament. But this tidy constitutional scheme was not destined to last. Between 1908 and the outbreak of the First World War, it gradually became more and more evident not that constitutionalism had replaced autocracy, but that army officers, engaging in successive *coups d'état*, had become the sole legatees of the Sultan's autocracy.

They were legatees whose power was more extensive, more ruthlessly used, and more remote from the governed than that of the sultans. The Young Turks were, equally, the legatees of the Westernizers who had set the Ottoman Empire on a new course during the nineteenth century. The officers of the Committee of Union and Progress were to discover very soon that their outlook gave offence to a great many people both in the capital and in the provinces. This discontent erupted on 13 April 1909 into mutiny by troops of the First Army Corps stationed in Istanbul who, led by officers who had risen from the ranks and who had not experienced the Westernization which the CUP officers had undergone in the European-style military schools and colleges, appealed to the Sultan to defend the religion against the unbelief emanating from Salonika. A First Secretary of the Palace recalled in his memoirs an exchange with one of these mutinous soldiers which vividly expresses the strong antagonistic feelings which the CUP's *coup d'état* aroused among the traditionally minded mass:

'They curse our religion and ridicule our morals. By God it's a sin!' I replied: 'Who curses and ridicules you? Look, my boy, I beat this man as well.' I pointed to Hasan Aga, one of the coffee-makers at the secretariat, who was with me that day. I was very fond of him but on occasion while fasting I had struck him. 'Hasan has to be beaten once in a while by his officers and superiors, especially in the army, so what is the matter?' He leaned his head toward me and said: 'My Aga, I would let you kill me. Go ahead, hit me between the eyes, it would not matter. But the ones who beat us are children, babies, and their mouths are full of curses. They curse our beliefs and our religion. Now that's a sin, isn't it?'[14]

The officers in Salonika determined to curb the rebellion which they felt sure had been instigated by the Sultan. A force drawn from the Third Army Corps stationed in Salonika marched on Istanbul and quelled the badly led and disorganized anti-CUP movement. It is interesting that in the course of the march on the capital, the officer leading the contingent, General Mahmud Shevket Pasha, should have confessed that:

I have got this far by telling the troops that the Padishah [i.e. the Sultan] and the nation are in danger . . . If the soldiers hear that we are planning a deposition they will rebel and we will be finished.[15]

But until they reached Istanbul, the soldiers never realized in support of what cause they were being marched. Once in control of the capital, Mahmud Shevket and the CUP leaders arranged that the Chamber of Deputies and the Senate should meet in a National Assembly (a body unknown to the 1876 Constitution then in force) which decided that Abd al-Hamid should be deposed. This was conveyed to him on 27 April. His brother Reshad succeeded him and was given the regnal title of Mehmed V, for was not Istanbul conquered in his day, just as his illustrious ancestor Mehmed II had conquered it in 1453?

The events of April 1909 made fully manifest and finally established that the Ottoman Sultan's traditional autocracy was ruined. But it shortly became clear that what replaced it was not a constitutional order, but simply another kind of autocracy—that of a group of military officers whose successful *coup d'état* had delivered the state into their ambitious and—as the sequel was to show—impetuous and unskilled hands.

From the very beginning there were different views on policy among the CUP officers and their supporters. These differences were no doubt related to their discordant ambitions and rivalries, and led, in 1911, to an open split in the ranks of the CUP. The faction in power, disliking the growing opposition to its rule, dissolved the Chamber of Deputies in January 1912. The general election which followed returned a Chamber in which all but six members were supporters of the Government. This was known as the 'big-stick' election. In July of that year some officers calling themselves the 'Saviour Officers' promoted disaffection in the army and brought down the CUP Government. A new Government took office according to the dictates of the conspirators. The Chamber so recently elected was now dissolved. In January 1913 CUP officers led by Enver, one of the principal figures of the 1908 *coup d'état*, invaded the Sublime Porte where the cabinet was meeting and forced Ministers to resign at gun-point, in the process, whether by accident or design, shooting the War Minister dead. Following this *coup d'état*, the CUP ruled the Empire without challenge. They took the Empire into the First World War on the side of the Central Powers, bringing it down to ruin, together with their own regime, in the débâcle of 1918. The ambition to establish a constitutional, parliamentary, and representative regime in the Empire which had fired so many officials and officers since the 1860s, thus ended in failure, the Sultan's stable autocracy being replaced by the unstable rule of ambitious and insecure officers.

These officers, and the officials who administered the departments of the modernized state, increasingly spoke a political language quite at odds with the habits and the universe of discourse of the traditional mass of the population. The attitude of this mass is perfectly exemplified by the words quoted above of the soldier who took part in the counter-*coups* of April 1909 in Istanbul. It may be seen from the widespread popular opposition to the Young Turks which became manifest in the months following the *coup d'état* of July 1908 in various, widely dispersed, centres of the Empire, but which the new rulers, controlling the army and the administration, were easily able to snuff out. Because the traditional rulers had, for all their autocracy, shared with the ruled a common universe of discourse, their rule had rested on the unspoken, implicit consent of their subjects. Again, for all its autocracy, the traditional order, by means of various contrivances which had been fashioned over time, had afforded a way whereby the various groups in society could be said to have enjoyed an informal representation. Namik Kemal had a point in claiming that the Janizary corps, which came to make and unmake sultans in Istanbul, served in its anarchic way to temper and moderate autocracy. The same point can be made, more convincingly, about the spokesmen of the *millets* whom traditional rulers recognized as speaking for their co-religionists. When these traditional devices were swept away in favour of Western-style representative bodies, the result was what might be called a crisis of representation in the Middle Eastern polity—a crisis aggravated by the forcible manner in which the cake of custom was broken. From 1876 onwards, it was necessarily by means of conspiracies and *coups d'état* that the establishment of constitutional and representative government was attempted. Thus, the very means which had to be employed rendered the end cumulatively more difficult of attainment.

Another Middle Eastern state which experienced an attempt to introduce constitutional and representative government was Persia. The country was in many ways different from its Middle Eastern neighbours. At the beginning of the sixteenth century, the newly established Safavid dynasty imposed Twelver Shiism as the official religion of the state, and in due course it became the religion of the overwhelming majority of the subjects. As has been said previously, Shiism, whatever its differences from the Sunni beliefs of the majority of the Muslims, inculcated the same quietism and passive obedience which the Sunni divines taught. The Safavids, being themselves Shiites and claiming collateral descent from the Imam Ali, *a fortiori* received the same passive obedience from their subjects.

Under the Safavid dynasty, as well as under the Qajars who ruled Persia from the end of the eighteenth century, Shiite divines continued to enjoy the same veneration from the faithful as hitherto. These divines, considered to be general agents of the Hidden Imam, had had a crucial role in keeping the Twelver community together when they lived, powerless, in a sea of Sunnis, under the sway of rulers whom they considered both illegitimate and oppressive. The divines attended to the affairs of the community and offered guidance to the faithful. Their attitude to a Shiite ruler was thus distinctly less subservient than that of Sunni divines towards Sunni rulers. The faithful paid tithes directly to them. This allowed them to be personally independent of the ruler, and to dispose, as well, of large funds which they could expend on religious education, attracting in the process a devoted following of students and disciples. Again, from the sixteenth century onwards, the main Shiite shrines, Najaf and Karbala, where the most eminent divines resided and taught, were under Ottoman control and thus beyond the reach of Persian rulers and any pressure and control they might wish to exercise.

These features of the relation between state and religion in Persia were to assume particular importance from the end of the nineteenth century when agitation by various groups in Persian society faced rulers with a difficult, not to say dangerous, situation. Persia did not, in the nineteenth century, experience modernization and Westernization to anything like the same extent as the Ottoman Empire and Egypt. During his long reign, which ran from 1848 to 1896, and particularly in its last two decades, Nasir al-Din Shah intermittently tried to introduce Western methods of administration. But these came to nothing for reasons explained by Lord Curzon in his magisterial work, *Persia and the Persian Question*, published in 1892. In the first place, the Shah had a 'childlike passion for novelty':

Just as, in the course of his European travels, he picked up a vast number of what appeared, to the Eastern mind, to be wonderful curiosities, but which since have been stacked in the various apartments of the palace, or put away and forgotten; so in the larger sphere of public policy and administration he is continuously taking up and pushing some new scheme or invention which when the caprice has been gratified, is neglected or allowed to expire . . . Nothing comes of any of these brilliant schemes, and the lumber rooms of the palace are not more full of broken mechanism and discarded bric-à-brac than are the pigeon-holes of the government bureaux of abortive reforms and dead fiascos.[16]

In the second place, as Curzon also pointed out, the Shah had little or no control over the execution of his projects once they had passed out of his hands into those of corrupt and self-seeking officials.

Like administrative modernization, the modernization of the armed forces, begun in a desultory and intermittent fashion from the 1830s onwards, was also undertaken by Nasir al-Din in the 1870s. With Russian advice and help, he established a Cossack Brigade with Russian officers in positions of command. The Persian state, in short, was much more ramshackle than the Ottoman state which had emerged from the *tanzimat*, and the Persian army could not play the same role as the Young Turk officers did in the Ottoman Empire.

The divines, on the other hand, came to present a powerful challenge to the regime during the last years of Nasir al-Din's reign, and in the reign of his successor Muzaffar al-Din Shah (1896–1907). Being greatly embarrassed for money and seeking to increase his revenues, Nasir al-Din had in 1890 awarded to a British syndicate a monopoly in the purchase, sale, and export of tobacco. The transaction aroused great opposition from those in Persia who had an interest in this commerce, as well as from members of the official class who thought that the issue could serve as a weapon to unseat the chief minister who was involved in the negotiation. The opposition was discreetly fuelled by Russia who feared that the monopoly would increase British penetration of, and influence in, Persia. What proved of much greater significance, then and later, was the determined and public opposition to the Shah's policy by the divines. Dislike of deepening European influence, and fear of increasing European encroachment, actuated the divines. The agitation began in Persia, but eventually came to be led by Mirza Hasan Shirazi, who was recognized as the most eminent divine of the age. Shirazi resided in Samarra, a Shiite shrine city in Ottoman Mesopotamia, and was thus outside the reach of the Shah's power. He asked the Shah to cancel the tobacco concession, but the Shah was not willing to do so. Shirazi, therefore, issued a *fatwa*, i.e. a legal opinion, which forbade Muslims to use tobacco so long as the concession was in force, and declaring that anyone who used it was in effect waging war on the Imam of the Age. The *fatwa* was extensively publicized by divines in Persia, and it created a great stir among the population who overwhelmingly obeyed Shirazi's injunction. The concession had eventually to be cancelled.

The victory by Shirazi and the divines over the Shah undoubtedly was a great blow to the Shah's prestige and the stability of his regime. It also gave divines a taste for political activism, and instilled in them a belief that they were the Shah's masters who could at any time raise the population against him.

The tendencies towards political activism were encouraged by the words and actions of the curious personage who went under the name of Jamal al-Din al-Afghani (1838–97). Generally taken for a Sunni in the Sunni world where he spent many years of his life, he was in reality a Persian-born Shiite. He cultivated a reputation of Islamic zeal, while he was in fact a secret unbeliever. In a life of wandering, he had come into contact with European political ideas and currents during sojourns in British India, London, Paris, and Moscow. He had learnt how to speak the language of religion and make use of the Islamic solidarity—still very strong among all classes in the Muslim world, for novel political purposes inspired by European example, very much in the manner expounded by his friend Malkam Khan in a passage quoted earlier in this Chapter. In the late 1880s Afghani was able to attach himself to Nasir al-Din while the monarch was on a visit to Europe, and accompanied him back to Tehran. There he seems to have become the focus of an agitation, with subversive and populist tones, against the Shah and his ministers. He was thereupon deported from Iran, and from Basra in Ottoman Mesopotamia wrote a long letter to Shirazi denouncing the Shah and urging him, for the sake of Islam, publicly to oppose the tobacco concession. His motive was probably revenge, and also support of Russian policy, since he had had long-standing relations with Russia.

From London, where he joined forces with Malkam Khan in attacking the Shah's despotism, and from Istanbul where Sultan Abd al-Hamid sought to use him as a propagandist for Islamic unity under his aegis, Afghani continued to direct a stream of populist subversion against Nasir al-Din. In 1896, one of his devoted Persian followers, inspired by his vengeful mentor, returned from Istanbul to Tehran where he assassinated Nasir al-Din.

The assassin was caught. At his interrogation he did not seek to deny that he had committed the deed. On the contrary he justified it by the great oppression which the Shah practised on his subjects, many of whom had in consequence fled their country:

After all these flocks of your sheep need a pasture in which they may graze, so that their milk may increase, and that they may be able both to suckle their young and to support your milking; not that you should constantly milk them as long as they have milk to give, and when they have none, should devour their flesh from their body.

What is remarkable here is not so much the oppression and the exactions described, which are real enough, but to which the subject had in the past been taught and accustomed passively to submit. What the assassin was now saying betokened a new, rebellious attitude to the ruler. The attitude was justified by a doctrine which the assassin, a simple unsophisticated man, could hardly have worked out for himself. When asked how he, anxious for the welfare of his country, could justify an act which might plunge it into disorder and confusion, he answered: 'Yes, that is true, but look at the history of the Franks: so long as blood was not shed to accomplish lofty aims, the object in view was not attained.' The interrogation also disclosed that the assassin considered his master to be the long-awaited Mahdi.

This episode sheds a great deal of light on the change in political attitudes which Afghani both signalizes and, through his teaching and his following, articulates, hastens, and spreads among the intellectual and official classes. What he portends is partiality for political activism, and a millenarian expectation that political action, and if necessary, violent action, will do away with obscurantism and oppression, and establish a new, more just, society. The Young Ottomans and the Young Turks were attracted to the constitutionalist strand in the European political tradition. Afghani and the multitude of Middle Eastern intellectuals who were to fall under the sway of his legend were, by contrast, greatly attracted to another strand in this tradition—the millenarian and revolutionary strand which increasingly came into prominence after 1789 in one European country after another. Afghani, Mazzini, Garibaldi, Bakunin, and Nechaev were not only contemporaries, but they also exhibit striking intellectual, and in some cases even temperamental, affinities.

Afghani's violent, millenarian activism came to Persians of the official and intellectual classes mixed with a great deal of other Western ideas about political freedom, popular sovereignty, and representative government. Such ideas came from the familiarity with European civilization which some Persians, mostly sent on diplomatic missions, came to acquire. Some of these recorded the lessons which they had derived from their contacts in books which disseminated knowledge about Western political ideas and practices among their readers, who were no doubt few, but who belonged to the crucially important official and intellectual classes. One of the earliest writers, however, neither lived in Persia nor wrote in Persian. Fath Ali Akhund-Zadah, also known as Akhundov (1812–78), was born in the Russian Caucasus, and spent his life there as an official in Russian employ. He looked upon himself as Persian and a Shiite, though writing in Turkish. He is best known for his plays, and for a project for reforming the Arabic alphabet.

Both the plays (which were translated into Persian and published in Tehran in 1871) and the reform project clearly indicate his impatience with Islam and its traditions. The plays are indeed scathing in their mockery of Muslim divines who are shown up to be hypocrites and impostors. Muslims, in his view, had to embrace Western civilization and constitutional government, both of which required Islam to be jettisoned:

O Iranians! If you could realize the advantage of liberty and human rights, you would have never tolerated slavery and humility; you would have studied the sciences, have set up societies, have united your powers, and have decided to save yourselves from the tyranny of the despot.[17]

In Europe, guided by such thinkers as Montesquieu, Rousseau, and Mirabeau, people had come to realize that to obtain constitutional government they had to rise against the oppressor, not to petition him to grant their rights.

A friend of Akhund-Zadah's was the diplomat Mustashar al-Dawlah (d. 1895). While serving as chargé d'affaires in Paris, he wrote and published a commentary on the French Constitution of 1791, *One Word*. The one word in question was law. It was law, and respect for law, which ensured justice, and it was justice which lay behind the unbelievable achievements of the West, which was a hundred times more advanced than Russia. Law, the secret of Western superiority, was enacted by a legislative assembly which emanated from the people whose will is, thus, the basis of all policy.

Another Westernizer was Mirza Husayn Khan Sipah Salar (1826–81) who served successively as diplomat, as Nasir al-Din's Chief Minister, and as Minister for Foreign Affairs. During a long stay in Istanbul, from 1858 to 1870, he came to admire the Ottoman *tanzimat* and the equality of treatment for all subjects which the reforms promised. He was aware of the criticisms made by Mustafa Fazil Pasha, and of the demands for constitutional, representative government. When he became Chief Minister, he persuaded Nasir al-Din to institute a cabinet system with ministers each heading a department of their own, and collectively deliberating on policy. He also convinced the Shah to visit Europe for the first time, in 1873, so that he might be persuaded to introduce Western methods in the government of the Empire.

It was probably as a result of his European travels and his increasing familiarity with European ideas that Nasir al-Din came to think it necessary to provide an outlet for the subjects' grievances, and to introduce the same kind of reform as that to which the Ottoman rescripts of 1839 and 1856 had aspired. Thus he ordered petition-boxes to be placed once a month in the public place in the larger towns. He himself kept the keys, and the boxes were to be opened in his presence.

But the Persian provincial governor was not to be got the better of by so transparent a machinery. He promptly ordered a watch to be kept on the boxes; and the bastinado was freely administered to any indiscreet person dropping a petition. Wherefore the petition-boxes remained permanently empty, and the Shah felicitated himself upon the singular contentment of his subjects.[18]

The Shah also issued a reform proclamation in May 1888. The duties of his station, he declared, made it incumbent upon him to ensure for his subjects 'the enjoyment of their rights and the preservation of their lives and property from molestation by oppressors'. The Shah affirmed that he would spare no effort 'to the end that the people, secure in their persons and property, shall, in perfect ease and tranquillity, employ themselves in affairs conducive to the spread of civilization and stability'. He proclaimed that all his subjects were 'free and independent as regards their persons and property', and that it was his 'will and pleasure' that they should employ their capital in productive enterprises promoting 'civilization and security'. This proclamation was accompanied by an order to provincial governors enjoining strict observance of the proclamation, and threatening that anyone infringing it would be 'punished in such a manner as to be the wonder of all beholders'.

By the end of the century, then, ideas of reform and modernization had obtained some currency among the official and intellectual classes and, as has just been seen, had even influenced the ruler himself. Also, there was a gradual increase in the number of those who were becoming acquainted with them, either through direct contact with Europe, or through Western literature, or more frequently through Persian newspapers opposed to the regime, published in London or Istanbul or Calcutta and which were introduced and circulated in Persia. The impact of this heady mixture of radical and sometimes even revolutionary ideas was to be apparent only a few years after Nasir al-Din's assassination.

Nasir al-Din was succeeded by his son Muzaffar al-Din, a much less forceful and decisive character than his father. Owing to the tobacco protest, and his father's assassination, the country of which he was now the ruler was in a troubled state. Other developments made the situation even more messy. Persian currency was based on silver, and silver at the end of the nineteenth century was a steadily depreciating metal. The standard of living, never very high, fell in consequence, and the poorer classes suffered increasing hardship. Government revenues, at the best of times prey to peculation and corruption, suffered likewise, while neither the Shah nor his ministers knew how to control and decrease government expenditure. In the absence of direct taxation, almost the only way open to the Government to increase its income was to raise the rate of customs duties and improve their collection. This was done, and a modernized customs administration set up which was organized and managed by European officials. But the new duties could not apply to European enterprises, since the hands of the Government were in this respect tied by various capitulation treaties. The higher rates, the discrimination in favour of foreign business, and the greater efficiency in the collection of duties aroused great discontent and opposition among the merchants of the bazaar. Their relations with the Government worsened greatly. The Government sought to impose its will by high-handed and repressive methods, and the discontent increased.

This discontent was channelled and mobilized by divines who had not forgotten the great victory they had achieved over the tobacco monopoly. Again, in the first few years of the new century, secret societies began to proliferate, devoted to constitutional reform and to spreading ideas such as Afghani's; these too found the discontented mercantile class in Tehran and other cities fertile ground for the spread of their ideas. The clash between the government and its now organized opponents came to a head in 1905–6. In 1905, a group of merchants took advantage of a long-standing Persian traditional practice whereby anyone fearing unjust treatment by the authorities could take sanctuary in a shrine where he expected to be immune from official molestation. The group, complaining of oppression by the Government, took sanctuary in a shrine near Tehran and demanded redress of their grievances. The Shah was abroad, and his heir, who was acting for him, pacified them by promising action on his father's return. The tension remained however. In December 1905 the Prime Minister accused some sugar merchants of overcharging and had them bastinadoed. This led to some 2,000 protesters again taking sanctuary in the shrine, where they remained until the following January, financed, provisioned, and supported by their friends and no doubt also by political figures who hoped to supplant the Prime Minister. The deadlock was broken by the Shah giving way and promising to accede to a demand made some months previously for the establishment of a 'house of justice'. The protesters were pacified and returned to Tehran. But when, in the months following, there was no sign of the promise being kept, tension rose again. Divines, preachers, and members of the secret societies publicly denounced the authorities for this broken promise. In June 1906 the Prime Minister attempted to expel from Tehran two influential preachers. There were riots and a strike of the bazaars. A large group of divines, merchants, and their sympathizers took sanctuary in Qum. And on 19 July, another group, which eventually increased to some 12,000, took sanctuary in the gardens of the British Legation. The Shah again retreated as he had done the previous January, dismissed his Prime Minister, and on 5 August he issued a rescript which now granted the more far-reaching demands which the leaders of the agitation were putting forward. The rescript set up a national consultative assembly, for which elections were ordered to be held. The elections took place at the beginning of September, and on 7 September the Shah opened the assembly. It quickly drafted and approved a Fundamental Law which the Shah signed on his deathbed on 30 December. This was supplemented by another Fundamental Law which was promulgated on 7 October 1907 by his son and successor, Muhammad Ali Shah.

The two sets of Fundamental Laws show clearly enough the influence of European ideas. But just as clearly they contain other notions quite incompatible with European notions. The Supplementary (and more important) Fundamental Laws declare that the powers of the Realm are all derived from the people (Article 26); and that sovereignty is a trust confided (as a Divine gift) by the people to the person of the King (Article 35). Ministers, it is also laid down, are responsible to both Chambers of the Assembly (Article 44). The earlier set of Fundamental Laws declare (in the second paragraph of the preamble) that the Shah was conferring on 'each individual of the people of our realm' the right to participate in the elections for the Assembly, which is declared (Article 2) 'to represent the whole of the people of Iran, who thus participate in the economic and political affairs of the country'. All of these provisions are clearly of Western provenance and represent the ideas and wishes of the Westernized intellectuals who took advantage of the agitation occasioned by economic discontent to put forward their own aims. Given the traditions and outlook of the peoples of Persia, we may very much doubt whether they shared these aims, or whether indeed they could at all comprehend them. The subjects of the Shah had never been accustomed to think of themselves as repositories of sovereignty, or that the power of their rulers really belonged to, and derived from them. These were, as in the upshot they proved to be, mere words.

Side by side with such democratic and populist statements, the Fundamental Laws also laid down that the ruler appoints and dismisses ministers and heads of government departments, and the supreme command of the army is vested in him (Supplementary Laws, Articles 46, 48, and 50). These provisions approximated more nearly to the realities of power in

Persia, but how they were to be reconciled with assertions of popular sovereignty and parliamentary accountability was by no means apparent. And of course they could not be reconciled. In so far as the Constitution of 1906–7 had reality, the reality derived from these provisions which codified age-old practice.

In the agitation which forced Muzaffar al-Din to concede an assembly and a constitution, the divines made common cause with the Westernizers against the monarch. Their triumph in the tobacco protest had given them a high idea of their power, and they no doubt relished using it once again in the cause of curbing the despotism of the Shah and his ministers, and to strike yet another blow against foreign influence, now exemplified by the European officials in the customs administration. In 1903, for instance, from the shrine cities in Mesopotamia, some of them addressed the Shah with harsh words to warn that the laws of the Prophet were being perverted, and that it was necessary to put a stop to liberty and the imitation of the Franks.

The Fundamental Laws enacted at the end of 1906, and the open dissemination of Westernized ideas hitherto debated in the decent obscurity of secret conclaves, showed that the ideas of the Franks were triumphing with a vengeance, and that to diminish the Shah's power in order to give a free hand to those bent upon spreading unbelief was a very bad bargain. The Supplementary Fundamental Laws passed ten months later attempted to banish the divines' misgivings. Article 1 establishes Twelver Shiism as the state religion and requires the Shah to profess and to promote it. Article 2 goes on to give the divines a central and privileged position in the constitution. It declares:

At no time must any enactment of the sacred National Consultative Assembly, established by the favour and assistance of His Holiness the Imam of the Age (may God hasten his glad Advent!), the favour of His Majesty the Shahinshah of Islam (may God immortalize his reign!) the care of the Proofs of Islam [i.e. the most eminent divines of the day] (may God multiply the like of them!) and the whole people of the Persian nation, be at variance with the sacred principles of

Islam or the laws established by His Holiness the Best of Mankind [i.e. the Prophet] (on whom and on whose household be the Blessings of God and his Peace!).

To make sure that this prohibition is enforced, the Article went on to institute a committee of five eminent divines to whom all proposed legislation would have to be submitted. This committee, the Article stipulated, would have the power to

reject and repudiate, wholly or in part, any such proposal which is at variance with the Sacred Laws of Islam, so that it shall not obtain the title of legality. In such matters the decision of this Ecclesiastical Committee shall be followed and obeyed, and this article shall continue unchanged until the appearance of His Holiness the Proof of the Age [i.e. the Hidden Imam whose reappearance will inaugurate the reign of justice and thus render the article redundant] (may God hasten his glad Advent!).

The Article did continue unchanged, at any rate until the 1906–7 constitution was swept away by Ayatollah Khomeini who toppled the monarchy in 1979. All through its existence, however, the Article remained a dead letter; the committee it prescribes was never set up, and no legislation was ever submitted to the Islamic criteria, agreement with which it enjoined. What the article does indicate, when it is placed side by side with the other articles which are of European inspiration, is how great a gap there was between the various groups who worked for a constitution, and how unworkable the constitution would have proved if it had been seriously attempted to work it. The estrangement which occurred very quickly after the establishment of the constitutional regime between the groups who had worked to bring it about, may be exemplified by the career and fate of a prominent Tehran divine, Sheikh Fazlallah Nuri. Starting by being a supporter of the Constitution, Fazlallah Nuri came to be persuaded that the Fundamental Laws proclaiming, for instance, the equality of all the inhabitants of Iran, went against the tenets of Islam which, most emphatically, do not consider Muslims and non-Muslims equal. He became a vocal opponent of the Assembly and its works, and sided with Muhammad Ali Shah, Muzaffar al-Din's son and successor, who, in June 1908, disbanded the Assembly and abolished the Constitution as contrary to Islam. When, the following year, Muhammad Ali Shah was overthrown, Fazl-allah Nuri was arrested, tried, and hanged.

Under the new constitutional regime, Persia experienced a great deal of disorder. The power and authority of the Shah was severely shaken; the Assembly, which, according to the constitution, was to legislate, control finances, oversee ministers, and approve their acts, was simply a collection of disparate individuals and groups who pulled in different ways, suspicious of one another, of the Shah and his ministers, and a prey to intrigues. Persia, again, was by no means a homogeneous country. It was held together by the passive obedience ingrained in the multitude by religion and tradition, and by fear of the capricious rule of the Shah, ramshackle and feeble enough though it was. Nothing new appeared to replace its badly dented image. There was, instead, much talk of freedom and popular sovereignty. Law and order began to suffer, and all kinds of ambitions and cupidities to rear their heads. In August 1907, the Prime Minister, who was opposed to the constitution and suspected of sympathies with Russia, was assassinated by a member of a secret society, who described himself as a banker and a *fida'i*, i.e. someone prepared to lay down his life for a holy cause. The term had sinister reverberations since it described the votaries of the sect of the Assassins who were sent by the Old Man of the Mountain on their deadly missions. In December, the new Prime Minister and his colleagues who took office after the August assassination, and who were in favour of the Assembly exercising its constitutional powers, were arrested at the Shah's order. But he was foiled in this attempt to reassert his authority by clamour in Tehran and the provinces which he was powerless to quell. The following February an attempt was made to assassinate him, but he escaped. In June the Cossack Brigade bombarded and cleared the Assembly. Popular leaders were arrested, and two of them were strangled. On 27 June, the Shah formally dissolved the Assembly and declared the constitution contrary to Islam and thus null and void.

But the Shah did not possess the power to control a country

now profoundly agitated, and where both new-fangled secret societies and tribal grandees saw an opportunity to seize power, either for ideological ends, or simply for the opportunities of enrichment which power can open up. The secret societies of Tabriz, the capital of Azerbaijan, with skilled and intrepid leaders, rose in rebellion and ousted the Shah's soldiers and officials. They held the city for nine months. Rasht in the north and Isfahan in the centre also became the focus of rebellions. From Isfahan a force of Bakhtiari tribesmen led by their tribal leaders marched on Tehran. They were joined by fighters who had set out from Rasht, and the joint force occupied Tehran on 13 July 1909. Three days later, Muhammad Ali, who had been given refuge in the Russian legation, abdicated and fled to Russia. The Assembly formally deposed him that same evening, and proclaimed his twelve-year-old son, Ahmad, as his successor.

Elections for the second session of the Assembly took place at the end of 1909. The reality of power, however, lay with the Bakhtiari chieftains whose followers were in occupation, and who neither understood nor cared for constitutions. They and their hangers-on formed the government. But the country was now in a parlous state, with various provinces in turmoil. In June 1911, the ex-Shah came back to Iran and attempted to regain his throne with the help of Lur and Turkoman tribes whom he had lured to his cause. He failed and had once more to flee. These events plunged the country into greater disorder. The disorder did not leave the Great Powers, who had interests in Persia, indifferent. An Anglo-Russian convention had, in 1907, reserved northern Persia for exclusively Russian, and southern Persia for exclusively British, influence. The disorders which followed Muhammad Ali's attempt to regain power were the cause or pretext of the Russians reinforcing the troops they had stationed in the north since the summer of 1909 when Tabriz was taken over by the secret societies. In the south, the British zone of influence, the Persian Government was unable to maintain security. Roads were at the mercy of tribes and highwaymen on the rampage, and British trade, conducted from India through the southern ports, suffered greatly. To maintain their position, the British landed troops in Bushire.

The occupation of large parts of the country by foreign troops was naturally not to the liking of the Government or the Assembly. But the situation could not be remedied unless Persia had an efficient and reliable armed force. Such a force had to be paid for. Persian public finances were, however, in disorder. To remedy this disorder, in May 1911 the Assembly appointed an American, W. Morgan Shuster, as Treasurer-General with very wide powers. Shuster's powers and his activities displeased the Russians, who were in effective occupation of northern Persia, while Russian officers commanded the Persian Cossack Brigade. At the end of November they sent an ultimatum demanding Shuster's dismissal. The Assembly demurred; and the Bakhtiari chieftains who had marched on Tehran to restore constitutional rule, and who dominated the Government, now forcibly disbanded the Assembly, and sent Shuster out of Persia.

When Ahmad Shah attained his majority in July 1914, the opportunity was taken to hold new elections. The Assembly which issued from these elections had a very short life. When the Ottomans joined the Central Powers in the war against the Allies, Persia became a battleground where the two sides vied for influence. The British had troops in the south, now greatly augmented by the Mesopotamian Expeditionary Force which had occupied Basra shortly after the outbreak of war; Russian forces were paramount in the north. The Ottomans and their German allies did their utmost to make things difficult for their enemies, and their efforts were facilitated by the natural dislike for both British and Russians among the official and intellectual classes. German activities in Tehran raised fears of a Russian occupation of Tehran, and in November 1915 the Assembly broke up, most of its members fleeing to Qum. The Assembly was not to meet again until 1921, when it found itself facing a vastly different situation which, however, was equally inimical to constitutional and representative government.

4
Turkey after 1919:
The Failure of Constitutionalism?

As in so many other parts of the world, the First World War was a watershed in the politics of the Middle East. The modernization which had been going on with varying intensity in many Middle Eastern countries during the nineteenth century, now became much more radical and its rhythm accelerated greatly. What took place in the Ottoman Empire, in Persia, and Egypt at the end of the war, and in the years which followed, destroyed what remained of the traditional manner of doing politics, accentuated the centralization and absolutism which had been the hallmark of the so-called reforms of the previous century, and at the same time brought forth new forces and tendencies which have, even now, by no means exhausted their potential. Just as the *tanzimat* in the end seriously destabilized Ottoman politics and society, so the forces released after 1918 likewise have destabilized the Middle East to the point that it has today become a byword for arbitrariness and disorder.

When war broke out in Europe in August 1914 it was by no means fated that the Ottomans should join in on the side of the Central Powers against Great Britain, France, and Russia. But the Young Turk Government was persuaded and manœuvred by forceful and clever German diplomacy, and through the impulsion of Enver Pasha, the Minister of War and the most powerful figure in the administration, into taking such a perilous gamble. The following November the die was cast.

The war proved disastrous for the Empire, for the Ottoman dynasty, and for the Young Turks whose leaders were utterly discredited and who fled the country shortly following the armistice of Mudros, signed on 31 October 1918. The armistice provided for total and unconditional surrender of all Ottoman armies on all fronts. British forces had already occupied Mesopotamia and the Levant and now occupied Istanbul and the Straits. Though the Sultan was still on the throne and his ministers still, even though nominally, at the head of the administration, the Empire was manifestly in dissolution. The British, the French, the Italians, the Greeks, and the Armenians (who had set up a republic in Russian Armenia following the downfall of the Tsarist regime) all coveted various parts of Ottoman territory. As mentioned in Chapter 3, Mustafa Kemal, a general who had gained a high reputation during the war, reacted strongly against these foreign interests and the passivity of the Sultan and his government in Istanbul, where they seemed to be under the thumbs of the occupation authorities. Mustafa Kemal had sound political judgement, and he was determined and decisive. When he found that almost no one was prepared to listen to his views in Istanbul, having been appointed as Inspector-General of the Ninth Army, he left Istanbul in May 1919 for Samsun on the Black Sea. During the same month, the Greeks, after obtaining the agreement of Lloyd George, Clemenceau, and Woodrow Wilson, invaded Anatolia in order to secure Izmir and its hinterland, an area of long-standing Greek settlement. Very quickly, without, and even against, the Government's order, Mustafa Kemal put himself at the head of resistance to this invasion. He resigned his commission and appealed for support from the population and the armed forces. His call was widely heeded, and by September he was President of the Standing Committee of the Society for the Defence of Rights in Anatolia and Rumelia. The following November, in what proved to be the last elections for the Ottoman Parliament, he was elected Deputy for Erzurum.

In December the Standing Committee established itself in Ankara, which was shortly to supersede Istanbul as capital of a new state called Turkey, made up of the predominantly Turkish-speaking areas of the Empire in Anatolia and Rumelia. This new state was now to be governed by the Turkish Grand

National Assembly which Mustafa Kemal opened in April 1920 and of which he was elected President. The Grand National Assembly, composed of some of the deputies who had been elected to the Istanbul Parliament, of notables, professionals, civil servants, army officers, men of religion, and tribal chiefs, assumed both legislative and executive powers. Under Mustafa Kemal's guidance, it organized the military and popular resistance to the Greeks in Western Anatolia and to the French in Cilicia. In August 1921, the Assembly appointed Mustafa Kemal commander-in-chief. Mustafa Kemal now led his forces against the Greeks and repulsed them in the long-drawn-out battle of Sakarya. Responding to this victory, the Assembly conferred on Mustafa Kemal the title '*Gazi*'. This Arabic word, meaning conqueror, is one which harked back to the earliest history of the Turks in Anatolia, when it was given to successful warriors who challenged Byzantine forces on the marches of their Empire. For the Turkish-speaking Muslims, and for the Muslim world at large, it signified that Mustafa Kemal was the latest champion of Islam in the centuries-old contest with Christendom. In 1935, pursuant to a law requiring Turks to adopt family names, Mustafa Kemal assumed the surname Atatürk (father-Turk), dropping the title *Gazi* and first name Mustafa, which both have Islamic connotations, and since then he has been universally known as Kemal Atatürk.

The Islamic aura which surrounded Mustafa Kemal was extremely useful in consolidating and extending his support among the population at large, at a time when the Ankara regime was beset by both external and internal threats, and by the existence of a rival government at Istanbul which, however ineffective, yet benefited from the legitimacy which inhered in the person of the Sultan, who represented in his own person the claim of an ancient dynasty to the unquestioning obedience of its subjects. Mustafa Kemal had, however, no intention of co-existing with the Sultan's regime, recognizing the Sultan as the legitimate ruler, or conceding that the Ottoman dynasty had any place in the governance of the new state. Whatever his disagreements with the Young Turk leaders after 1908, he was after all imbued with Young Turk ideas. He may even be said to have taken these ideas to their logical conclusion. The Young Turks had been profoundly opposed to the Sultan's autocracy, and they also believed that the salvation of the state lay in the adoption of European political institutions and social customs. Fully sharing such ideas, Mustafa Kemal was now even more persuaded, in view of the Sultan's passive acquiescence in the foreign occupation and his antagonism to the resistance movement in Anatolia, that the *ancien régime* had to go. But he was extremely circumspect in giving voice publicly to these ideas. In a secret debate which took place on 25 September 1920, he took exception to the idea, mooted by some members of the Grand National Assembly, that the activity and existence of the Assembly were temporary, to last only until the Sultan had been liberated from foreign domination. He declared that at a time when they were struggling to preserve the independence of the country, 'an excessive preoccupation' with the sultanate and caliphate was bound to be harmful:

The highest interests require that no mention should be made of them for the present. If the purpose is to express and to confirm continued attachment and loyalty to the present Caliph and Sultan, then it should be known that he is a traitor. He is an instrument used by the enemy against our country and our people. As long as he is called Caliph or Sultan the people will obey his orders and will be bound to carry out the designs of the enemy. A traitor or a person prevented from exercising the power and the authority of his calling cannot be either Sultan or Caliph.

Mustafa Kemal's words were both clear and clear-cut. Thus in January 1921 when the Grand National Assembly enacted a Law of Fundamental Organization (as mentioned in Chapter 3), it left on one side the issue of the Sultanate. Article 1 declared: 'Sovereignty belongs unconditionally to the nation. The system of administration is based on the principle that the people personally and effectively direct their own destinies.' The Ottoman Constitution of 1876 had declared sovereignty to 'belong' to the eldest descendant of the Ottoman house. This new instrument declares it to belong unconditionally to the nation. Here indeed was a revolution which would prove far-reaching, even though its immediate consequences for those who were said to be its unconditional beneficiaries were not easily discerned.

If sovereignty belonged unconditionally to the nation, what place did the Sultan have in its governance? The issue was decided soon afterwards. After the Greek rout and the occupation of Izmir by the forces of the Grand National Assembly in September 1922, a peace conference was called to settle the Greco-Turkish hostilities and other pending Middle Eastern issues. To this conference, due to meet at Lausanne, both the Istanbul and the Ankara governments were invited. This was in no way acceptable to Mustafa Kemal, and the problem of the sultanate was thus brought to a head.

On 1 November the Grand National Assembly debated the issue. Mustafa Kemal proposed the abolition of the sultanate in the face of much opposition, particularly from men of religion. In the course of this debate he declared:

Sovereignty and sultanate are not given to anyone by anyone because scholarship proves that they should be; or through discussion and debate. Sovereignty and sultanate are taken by strength, by power and by force. It was by force that the sons of Osman seized the sovereignty and sultanate of the Turkish nation; they have maintained this usurpation for six centuries. Now the Turkish nation has rebelled, has put a stop to these usurpers and has effectively taken sovereignty and sultanate into its own hands.

The Grand National Assembly then proceeded to pass a retroactive resolution to the effect that 'the Turkish people consider that the former government in Istanbul, resting in the sovereignty of an individual, ceased to exist on 16 March 1920 [i.e. when Allied troops occupied Istanbul], and passed for ever into history'. A few days afterwards, the Sultan, Mehmet VI Vahideddin, left Istanbul and fled to Malta on board a British warship.

The Grand National Assembly passed a second resolution that day. This other resolution treated the caliphate as an office separate from the sultanate. The caliphate, the resolution

recognized, inhered in the Ottoman dynasty, but it declared
that it rested on the Turkish state, and that the Assembly would
choose 'that member of the Ottoman house who was in bearing
and character most worthy and fitting'. A cousin of the deposed
Sultan, Abd al-Majid, was chosen.

This step was, in Islamic terms, even more revolutionary
than the affirmation that sovereignty belonged unconditionally
to the people. In effect, this step meant that the office of
Caliph, the highest politico-religious office in Islam, was now,
by fiat of an assembly, to be stripped of all power. The Caliph
was to become a mere spiritual head, a kind of Pope. This was
signalized by the instruction the new Caliph was given to entitle
himself Caliph of the Muslims. But the idea of a merely spiritual
office was alien alike to Islamic thought and practice. The
Ottoman sultans had claimed the Caliphate because they were
the rulers of an extensive and powerful state. It was because
they were sultans that they were recognized by most of the
Sunnis as Caliphs, and not the other way round. What, in any
case, was this new-fangled Caliph, subordinated to a popular
assembly, to do? If he carried the title, he and many others
would go on believing that his position was exactly the same as
that of his predecessors. This ambiguous situation in which two
authorities were potential rivals, each residing in a different
city, could not possibly endure. Mustafa Kemal, with his usual
clear-headed decisiveness, put an end to it promptly and
expeditiously. On 13 October 1923 the Assembly voted that
Ankara would replace Istanbul as the capital. At the end of the
same month, having prepared the ground and neutralized the
opposition to so momentous a step, Mustafa introduced in the
Assembly an amendment to the Constitution declaring that 'the
form of government of the state of Turkey is a Republic'. After
a long debate which showed how many members found this
novel proposal strange and shocking, the amendment was
passed by 158 votes, the rest of the 286 deputies abstaining.
Mustafa Kemal, hitherto President of the Grand National
Assembly, was, at the same sitting, elected President of the
Republic.

There remained the Caliph, sitting high and dry in his palace at Istanbul. With the removal of the capital to Ankara, and with the declaration of a republic, his position, from being equivocal, was now becoming intolerable. Simply by being there, he might become the focus of Muslim opposition to the new order inside Turkey, and encourage Muslims abroad to bring pressure to bear on Mustafa Kemal in order to deflect him from the changes which he absolutely believed were necessary. Now that the caliphate had been stripped of all political power, it could act as the symbol and standard-bearer of Islam world-wide. For it to do so, the Caliph's position—so some eminent Indian Muslims publicly argued in November 1923—had to be such as to command the confidence and esteem of Muslim nations. This attempt to bring pressure on the Ankara government decided Mustafa Kemal finally to have done with the Ottoman dynasty. Aware of the opposition this would arouse, Mustafa Kemal again carefully prepared the ground, and on 3 March 1924 motions were put before the Assembly to depose the Caliph, abolish the Caliphate, and banish from Turkey all members of the Ottoman dynasty. The following morning Abd al-Majid was driven to a suburban station outside Istanbul and put on the Orient Express bound for Europe. Other measures swiftly followed, designed to break the hold of Islam and its traditions on the population. The office of *Sheikh al-islam* the highest religio-legal authority in the Empire, the ministry of the *shari'a*, that of Pious Foundations, and the *shari'a* courts were all in turn abolished. So likewise were religious schools and colleges.

The abolition of the Ottoman caliphate came as a shock to the Muslim world, even though a caliphate divorced from political power was pretty meaningless in Islamic terms. The shock was made even more grievous by the wholesale abolition of religious institutions in Turkey which accompanied it. Could the Muslim world do anything to replace what the Turks had destroyed? Long before the demise of the caliphate was even conceivable, in the last two decades before 1914, there had been, here and there, talk about the Ottomans being divested of the caliphate, on the score that a legitimate Caliph should, according to the doctors of the law, be descended from Quraysh, the Prophet's tribe, and the Ottoman dynasty manifestly was not. Abbas Hilmi, the Egyptian Khedive, who was an ambitious man and a lover of intrigue, may have helped to spread it, either in the hope of increasing his own power at the Ottomans' expense, or as a way of blackmailing the Sultan into settling various, mostly financial, disputes in his favour. The idea was put about that the caliphate should devolve upon the Sherif of Mecca, certainly a descendant of the Prophet, but also holding office at the pleasure of a Sultan, one of whose titles was indeed that of Servant of the Two Noble Shrines [i.e. Mecca and Medina]. Kawakibi, who was mentioned in the previous Chapter, published articles in an Arabic periodical, later assembled in a book, which argued that the caliphate belonged to Quraysh, and that it should be held by the Sherif of Mecca. The Arabian caliphate for which Kawakibi argued would be a spiritual one, making Mecca the spiritual and religious centre of Islam and, so he hinted, would be under Abbas Hilmi's protection.

About two years after the outbreak of war in 1914, Husayn, the then Sherif of Mecca, rebelled against the Ottomans. In the secret exchanges with the British which preceded the rebellion, Lord Kitchener, the War Secretary, dangled the prospect of the Sherif becoming Caliph. Kitchener could not have been offering Husayn rule over the extensive domains of the Sultan. The prospect of a Qurayshite caliphate which he held out to the Sherif had a religious and not a political character. Between Kitchener and Husayn, however, talk about the caliphate could only be talk at cross-purposes, since for Husayn the caliphate meant what it had always traditionally meant, namely temporal, and only in consequence spiritual, dominion. The prospect which Kitchener held out to Husayn proved illusory and, in spite of his great ambitions, Husayn ended the war as a mere King of the Hijaz.

Even a caliphate so dramatically diminished by the actions of the Turkish Grand National Assembly had its great attractions, to judge from the eagerness with which some contenders sought the honour. The King of the Hijaz had himself proclaimed Caliph, it being claimed on his behalf that he had the suffrages of the faithful in Iraq and Transjordan where two of his sons reigned, and perhaps also elsewhere in the Levant where the Sherifians had followers. But no one took seriously Husayn's caliphate and it quickly lapsed into obscurity. The King of Egypt, Fuad, descendant of Muhammad Ali and uncle, as it happened, of Abbas Hilmi, also considered himself a fit candidate for the caliphate. Immediately following Abd al-Majid's deposition, Egyptian divines, obviously acting under his inspiration, immediately called for a congress to meet in Cairo and discuss the issues which this deposition raised for the Muslim world. This was a transparent device to create a movement in support of Fuad's Caliphate. But support was by no means forthcoming. The congress was meant to meet in 1925, but was postponed. When it met in May 1926, it became clear that there could be no question of electing Fuad Caliph. No Muslim country was prepared to recognize Fuad's primacy, religious or otherwise. The congress, which took two years to assemble, lasted barely a week and held only four meetings. A close Egyptian observer wrote at the time:

The delegations [to the congress] said that a caliphate was obligatory! They then pointed out the impossibility at present of establishing it among the Muslims! Finally they decided to found branches of the congress in different Islamic countries so as to prepare further successive congresses, as need be, in order to decide the issue of the caliphate![1]

Faruq, Fuad's son and successor, also had the ambition to become Caliph. Shortly after he ascended the throne an agitation was got up in Egypt, in 1937–8, whereby he was publicly hailed by the crowd as Commander of the Faithful. And even as late as 1952, only a few months before Faruq's deposition following the military *coup d'état* in July, a committee, clearly inspired and instigated by the royal palace, endeavoured to establish a Qurayshite genealogy for the King. It was here, in this committee, 'that the venerable caliphate suffered its last agony, and expired hanging from a fake genealogical tree'.[2]

These failed attempts to elect a new Caliph show that the idea of the caliphate no longer had any value in the market-place of political ideas. On the one hand, in the aftermath of the First World War a number of Middle Eastern states had come into being whose new rulers were not attracted by, and had no allegiance to, the traditional ideas embodied in the notion of the caliphate. And on the other, Muslim political thought could not find within itself resources to modify the idea and to make it workable under modern conditions. It is significant that one Egyptian divine and judge in the religious courts, Ali Abd al-Raziq, published in 1925 a tract on *Islam and the Foundations of Authority* in which he went so far as to argue that the caliphate was not, properly, part of Islam, and its institution was not therefore a religious duty: a caliph, he argued, was no more than a king. The argument went a great deal against traditional Islamic presuppositions and it was denounced by the *ulema* as heretical. Ali Abd al-Raziq was summoned before a tribunal of *ulema*, tried and convicted of holding unsound opinions, and deprived of his status as doctor of religion and judge. In the writings of the traditionalist divines, however, the Caliph remained what he always had been: the highest and sole legitimate ruler of the Muslim community from whom all authority flowed. This notion had long ceased to reflect or describe the realities of Muslim politics. As was seen in Chapter 1, the response to the jurists had been, on the one hand, to ordain obedience to any ruler able to seize power, and on the other to maintain the caliphate as a bookish, unattainable and hence only a pale, ineffective ideal.

To palliate its manifest and utter failure, the Cairo caliphate congress suggested, as has been seen, that congresses should be held in future 'in order to decide the issue of the caliphate'. The idea of a congress or conference to debate and decide on Islamic issues is inspired by Western example. Before contact with Europe, it had never occurred to anyone in the Muslim world to convene such assemblies. The caliphate congress is a very early example of a practice which has now become quite common. However, the future caliphate congresses envisaged in the congress of 1926 did not materialize. Failing a caliph who might be looked up to as the symbol and the representative of the Islamic *umma*, in some sense the guardian—and perhaps also the reformer—of the Law, some Muslim thinkers thought that learned and authoritative Muslim jurists, meeting periodically, might fill the place of the caliphate and might, over time, make changes in the corpus of Muslim law to take account of problems and tendencies emerging in the modern world. This idea also came to nothing.

The affirmation that sovereignty unconditionally belongs to the nation, the deposition of the Sultan, the proclamation of a republic, and the abolition of the Ottoman caliphate—steps successively taken within just over three years—may be said to have constituted a revolutionary transformation: the Ottoman sultanate was now a Turkish republic. The driving force behind this radical change, which was effected in the face of much determined resistance, was entirely Mustafa Kemal. What, however, did this juridical change substantively signify for political and social life? As was to be seen in the years which followed, in fact a great deal. For Mustafa Kemal, these constitutional changes were only the prelude of a sustained attempt to change out of recognition the character of the people he was to govern until his death in November 1938. Shortly after the abolition of the caliphate, the Grand National Assembly adopted a new Constitution on 20 April 1924, which in its first article affirmed that 'The Turkish State is a Republic', while its third article repeated the declaration which figured at the head of the 1921 constitution, namely that 'Sovereignty belongs unconditionally to the nation'. Article 2, however, laid down that the religion of Turkey was Islam, and other articles obliged the Grand National Assembly to see to the implementation of the *shari'a*, and yet others incorporated a mention of Allah in the oath of office taken by the President of the Republic and members of the Grand National Assembly. These provisions figured in the Constitution not because Mustafa Kemal advocated or was attached to them. Rather it was because to eliminate all reference to Islam would have provoked too great an opposition. As it was, opposition to the abolition of the caliphate, the office of *sheikh al-Islam*, and the Department of Pious Foundations was not confined to speakers in the Assembly. In February 1925, a rebellion broke out in the eastern provinces and spread to the south-east, the leader of which was Sheikh Sa'id, hereditary chief of the Naqshbendi order of dervishes. Mustafa Kemal acted swiftly. A 'Law for the Maintenance of Order' was passed which gave sweeping powers to the Government for two years. It was renewed in 1927 for another two years. 'Independence tribunals' with summary powers of execution were set up. Swift military action crushed the uprising, and the Sheikh, together with forty-six of his followers, were sentenced to death at the end of June and executed the following day.

The rebellion and its defeat was followed in September by the abolishing of dervish orders and monasteries, and closing down of tombs of holy men—saint-worship being considered a harmful superstition. A decree also abolished the use of religious titles and distinctive clothes associated with them. In a speech a few weeks before, Mustafa Kemal declared that the Turkish Republic could not harbour dervishes and their orders:

The first *tarikat* [i.e. order] is the *tarikat* of civilization. If we want to be men we must carry out the dictates and requirements of civilization . . . We draw our strength from civilization, scholarship, and science and are guided by them. We do not accept anything else. The [dervish monasteries] want to make ecstatic fools of the people, but the people have decided not to become ecstatic fools.

Mustafa Kemal took other action in pursuance of his aim to eradicate the harmful legacy of Islam, and securely establish European civilization in Turkey. Also following the rebellion, he began openly to advocate the abandonment of distinctive Islamic modes of dress, and in particular the covering of the head, and the use of such distinctive headgear as the fez. Prior to that, in the spring of 1925, a peaked cap was introduced in the army, and some deputies began to appear bareheaded in the Assembly. In speeches in August 1925 he attacked traditional dress and headgear:

We must become civilized men from every point of view . . . our thinking and our mentality will become civilized from head to foot . . . The nation should realize clearly that civilization is a powerful fire which burns and destroys those who disregard it. We shall acquire, keep, and finally improve the place we deserve in the civilized family to which we belong.

Civilized international forms of dress are a worthy costume for our nation. We shall wear shoes or boots on our feet, then trousers, waistcoats, shirts, ties, jackets, and, naturally to crown it all, we shall wear brimmed headgear. I want to say clearly that this brimmed headgear is called a hat.

On 25 November 1925 the Assembly (with one deputy dissenting) adopted a law making it illegal for anyone to wear a fez or the brimless fur hat known as the *kalpak*. This anti-Islamic campaign culminated with the amendment in 1928 of Article 2 of the 1924 Constitution, removing the stipulation that Islam was the official religion of Turkey. And in 1937 two other amendments reinforced the anti-Islamic character of the Constitution. Article 75 had provided for freedom of religious convictions and philosophical beliefs, and for the right to belong to different sects: the word 'sect' was now deleted. Again Article 2, which had declared that Islam was the official religion of the state, now contained a recital of the tenets of Kemalism which became the official ideology of Turkey: the Turkish State, the article *inter alia* declares, is secular.

The prohibitions explicitly or implicitly laid down in the Constitution of 1924 and its amendments were supplemented by other laws emphasizing the rejection of Islam and its culture as a force in Turkish society: namely, a law for the unification of education, which abolished religious schools; a law requiring that marriages should be solemnized under the civil code enacted, on the Swiss model, in 1926; a law requiring the replacement of Islamic time and calendar systems by European ones, enacted in 1925; another requiring the adoption of international numerals, adopted in 1928; another law of the same year substituting the Latin alphabet in place of the Arabic alphabet, with all its Islamic associations; another law of 1926 making legal the consumption of alcohol and spirits by Muslims, and a law of 1934 which banned certain garments with Islamic associations from being worn. The most important of these provisions and prohibitions were reiterated and entrenched in the Constitution of 1961, of which Article 2 affirmed the secular character of the state, and Article 153 listed the laws just described and declared that no provision in the constitution shall be 'construed or interpreted as rendering unconstitutional [these] . . . Reform Laws which aim at raising Turkish society to the level of contemporary civilization and at safeguarding the secular character of the Republic'. Article 174 of the 1982 Constitution repeats word for word the Article of the earlier constitution. But this latest Constitution seems to go even further in imposing an absolute ban on any reflection of religious concerns in social and political life. Its Article 24 declares that 'Everyone has the right to freedom of conscience, religious belief, and conviction', but goes on to stipulate that 'No one shall be allowed to exploit or abuse religion or religious feelings, or things held sacred by religion, in any manner whatsoever, for the purpose of personal or political influence, or for even partially basing the fundamental, social, economic, political, and legal order of the State on religious tenets'.

These categorical and draconian prohibitions reinforced and repeated from decade to decade, from one constitution to another, are indication enough that what obtains here is not the state of affairs encountered in modern European polities and usually described by the phrase, 'a free church in a free state'. What seems unmistakably the case here is that, as in the era of the *tanzimat*, there is a gap, a divorce, between the outlook, the beliefs, the ideals of the rulers and those which the ruled, or at any rate very large numbers of them, persist in preferring. In a viable, workable constitutional order the officially defined and recognized norms of social and political life do not go against the grain of society. If they do, rulers have to keep the ruled at arm's length; they have to be always on guard lest society at large should, by passive opposition, at length undermine the norms in which the rulers believe; or else that the gap between its behaviour and the norms in force would periodically lead to great political tension, putting at risk the very political order. The question thus has to arise of whether in a country like Turkey constitutionalism and the realization of Mustafa Kemal's vision were compatible.

The political history of Turkey under Mustafa Kemal and, subsequently, Ismet Inönü, his successor in the presidency of the Republic, who had been perhaps the most prominent figure after him in the anti-Greek struggle and in the Ankara government, shows that in fact there was little compatibility. The Grand National Assembly was the outcome of Mustafa Kemal's initiative in raising the standard of resistance to foreign invasion. He acted, and was seen by the mass of the people to act, in default of a defeatist government in Istanbul. In their eyes the issue was simple: whether the Muslims were to be invaded by, and subjected to, Christians who not long ago were the Sultan's subjects. Hence the elections of 1920 produced an Assembly the overwhelming majority in which were supporters of the *Gazi* and the armed resistance which he was organizing. When, having resigned from the Ottoman Army, Mustafa Kemal had started to organize the resistance, he did so as the Chairman of the Association for the Defence of the Rights of Anatolia and Rumelia. Before the elections to the second Assembly took place, Mustafa Kemal took steps to transform the Association into a party—the People's Party, later to be called the Republican People's Party. In an interview with newspaper editors in January 1923, he declared: 'When I say that I shall form a party under the name of People's Party, it should not be thought that my aim is to work for the prosperity of any one class of the people. The party's programme will aim at the prosperity and happiness of the whole nation.' The Party, then, was meant to be quite different from the pre-war Committee of Union and Progress and its (short-lived) rival, the Liberal Union Party. These were factions which had developed out of the conspiratorial activities of Young Turk officers. As such, they could not be said to represent, in any sense of the term, a particular interest or class in Ottoman society. They were simply groups of officers and their civilian hangers-on who sought to gain control of the state and its institutions through the control and (illegitimate) use of military power. The People's Party was meant by its founder to be an instrument for the political control of the masses, and as a transmission-belt, auxiliary to the administration, the purpose of which was to promote Mustafa Kemal's secularist project. As he put it in a speech: 'The People's Party will be a school for the political education of our people'.

As has been said above, Mustafa Kemal's secularist project was not without its opponents—opponents, like Sheikh Saʿid, who were ready to resort to rebellion. But there were also other opponents from among his earliest supporters, who objected to the manner in which he was monopolizing power, and to the autocratic character of the government over which Ismet Inönü presided as Prime Minister. A few of these opponents left the People's Party towards the end of 1924 to found a Progressive Republican Party. In response to criticism, therefore, Mustafa Kemal replaced Ismet Inönü by an ex-officer, Ali Fethi Okyar, supposed to be more liberal than his predecessor. However, with the political traditions of a country like Turkey, Mustafa Kemal's action could be easily misinterpreted. Allowing a new party and appointing a new prime minister could easily be taken to mean that the ruler, a *Gazi* who was like a sultan, who had actually toppled a sultan, was indicating that he had disgraced his former favourites in favour of their rivals. Alternatively, if a party had come into being without the ruler's approval, then it must be that his authority was on the wane, and that his opponents were the coming power and should thus be supported. Thus, either way, its very existence attracted to the Progressive Republican Party supporters who, anyway, had never been reconciled to the secularism of the regime. The formation of this Party was shortly followed by Sheikh Saʿid's rebellion which Mustafa Kemal took very seriously. Fethi Okyar was dismissed at the end of March 1925 and Ismet Inönü was once more Prime Minister. The Party was dissolved by decree the following June.

Having, between 1925 and 1929, crushed with heavy-handed and draconian methods, all opposition to the regime, Mustafa Kemal was now minded to license another opposition party. In August 1930, Fethi Okyar, encouraged by the President, obtained permission to set up the Free Republican Party, whose programme was an implicit critique of the autocracy of the regime, and Ismet Inönü, still Prime Minister. But again, the appearance of an alternative party led to the occasionally violent expression of long pent-up opposition to the regime and the rough methods it was inclined to employ. In November, Fethi dissolved the Party, 'because struggle against the *Gazi* was impossible'.

Under Kemal Atatürk until his death, and under Ismet İnönü from 1938 to 1946, the Republican People's Party remained the sole legally recognized party. It was not the kind of party familiar in constitutional and representative regimes, namely a group representing a particular interest or espousing a particular policy, and competing for the favours of the electorate. It was meant rather to indoctrinate the population in Kemalist ideology, the tenets of which—republicanism, nationalism, populism, etatism, secularism, and reformism—were in 1937 enumerated in Article 2, and thus made part of the Constitution. The Party established a network of People's Houses in cities and larger towns, and People's Rooms in smaller towns and villages through which the population could be indoctrinated and politically mobilized. The distinction between the Party and the Government was so tenuous as to be almost invisible. In 1935 the Minister of the Interior and the Secretary-General of the Party were one and the same; likewise the provincial governor and the chairman of the Party organization in the province. The identity between Party and Government became ingrained in the popular mind. A writer discussing Turkish politics long after the Republican People's Party had lost its monopoly, cited a Turkish schoolboy's suggestion which was both ingenious and revealing. The schoolboy, casting about for pure Turkish words to replace words derived from Arabic, came up with the idea that the word *parti* (adopted in modern Turkish to replace *firka* which had been the common usage, but which was tainted with an Arabic origin) was the pure Turkish equivalent of the Arabic-derived word for government, *hükumet*, also hitherto in common use.[3]

Party-state autocracy continued until the end of the Second World War, during most of which Turkey had succeeded in remaining neutral. However, when the result was no longer in doubt, Turkey declared war on Germany and its allies, in

February 1945, and thus became eligible to join the newly formed United Nations, to the Charter of which it subscribed. The end of the war saw the international situation changed out of all recognition. Turkey's powerful neighbour, the Soviet Union, had become even more powerful. With all the familiar landmarks of the European state system swept away, the only shield against Soviet ambitions seemed to be United States protection. Not only was this new superpower itself democratic in its governance, but its leaders also believed that democracy should be promoted, world-wide—a belief mirrored in the rhetoric of the Atlantic Charter which Roosevelt and Churchill published in the middle of the war, and in the language of the UN Charter. This may have been the reason why the President, Inönü, found it prudent, and perhaps also necessary, to adopt this language in public pronouncements, and to hold out the prospect that party-state autocracy would no longer bear so heavily on the country, and that 'democratic principles' would be the rule even more than in the past! The first fruits of this policy were seen when the Government decided, in June 1945, that it would not support official candidates for some by-elections then pending. At about the same time a group of deputies belonging, of course, to the Party, publicly demanded that the commitment to democracy entered into through the signing of the UN Charter should be shown by means of concrete measures. The Party decided to expel two of the leaders of the group, Adnan Menderes, a lawyer and land-owner, and Fuad Köprülü, a well-known and distinguished scholar. Another deputy, Refik Koraltan, a lawyer and former official, publicly defended his two colleagues, whereupon he too was expelled from the Party. A fourth leader of the dissidents, Celal Bayar, who had played a part in the war of independence under Mustafa Kemal, and served as Prime Minister in 1937–9, resigned from the Party in support of his expelled colleagues.

This, then, was a public schism within the Party, and the schism led to the formation of the Democratic Party in January 1946. It was led by the four figures who had been recently expelled. It is not known why exactly Inönü allowed the formation of an opposition party. Was it the same kind of action which Mustafa Kemal had taken in 1924 and 1929, which he had promptly revoked, and which had in no way shaken his hold on power? Was it the desire to prove to the world that the regime was indeed democratic? However, as has been said above, an act of this kind on the part of an all-powerful leader, in a polity like Turkey, was bound to be understood either as showing that the leader was not all-powerful, or that he was abandoning his old, in favour of new, supporters. Thus, the Democratic Party, simply by virtue of having been allowed to exist, attracted support which, in a snowball effect, went on increasing.

The new Party, formed in January, had to fight elections which suddenly the following April, the Republican People's Party decided would take place in July. The governing Party's apparatus was still in place and functioning, the nexus between party workers and officials was still in existence, and voters could not bring themselves to abandon habits inculcated in two decades of control and regimentation. Out of 465 seats the Democrats contested 273 and won 61. This was not at all a negligible accomplishment for a party which had not been heard of six months before. Now it was recognized as legitimate and it had a presence in the Assembly large enough to question and contest the actions of the administration, and to publish criticisms and denunciations. Events which followed the 1946 elections served to make the Democratic Party's challenge more formidable. When the Assembly convened after the elections, relations between majority and minority became increasingly acrimonious. Inönü thought it necessary to intervene by convening a series of meetings with the Prime Minister (who was the chairman of the ruling party) and the Democrats. He then published a declaration to the effect that as President he had to be above politics and even-handed as between government and opposition. To the stalwarts of the Republican People's Party this could not but be disconcerting and demoralizing. Changes of government, with 'liberals' in the Party taking up important posts, increasingly gave the impression of a party and an administration adopting the rhetoric and the posture of the opposition. Thus, in January 1949 the Prime Minister of the day, Professor Günaltay, declared that he would sincerely work to establish democracy, and that the rules of the Western democracies would now serve as a model. This too must have added to the demoralization of the Party followers. Again, martial law, which had operated during the war years, was lifted in December 1947. This gave much greater scope for criticism of the Government and the Party, which, feeding upon itself, increased in scope and volume. In retrospect one can see in the period 1946–50 a veritable rout in the making for the Republican People's Party.

The rout became fully manifest in the elections of 14 May 1950. The Democrats came out with 420 members and Atatürk's party with only 63. In November 1945 Inönü had announced a change from indirect to direct elections. Men and women over the age of 22 now could choose their own representative. In the 1950 elections there were about a million voters, and of these over 88 per cent cast, in some fashion or another, their suffrages. It is not to be supposed that this mass of voters predominantly rural and of whom a very high proportion—perhaps two-thirds—would have been illiterate, had coolly and judiciously considered the pros and cons of the Republican and Democrat programmes put before them by city-bred, Westernized *Efendis*, with whom they had little in common. To explain such an astounding phenomenon one would have to say that the modest snowball which had begun forming in 1946 had kept on growing and gathering pace, until it had all but crushed those who strove, somewhat maladroitly, to arrest its progress. What might have helped to increase the size of the snowball was, possibly, the attitude of the bureaucracy. It has been said that its neutrality in the electoral battle was secured by the Democrats promising, in case of victory, not to rake over the past. Neutrality was essential in order to convey to voters that the ruling Party no longer enjoyed the support of the administration, so long taken for granted. If the bureaucracy was really deserting the Republicans, then this would automatically redound to the Democrats' prestige and encourage voters to flock to them.[4]

Adnan Menderes became Prime Minister and Celal Bayar President. The country which they were now to rule for ten years was in many ways trickier and more difficult to govern than before, and would become even more so as a result of their own activities. Since Mustafa Kemal took power, the population had been steadily rising, and the rise was also gradually accelerating. From 13.5 million in 1927, it had become nearly 21 million in 1950. The country, when the Ankara Government took over, never by any means rich, had sustained great losses during the First World War and the various hostilities which followed. As part of the settlement of the dispute with Greece, an exchange of populations was agreed: of Greeks from Anatolia and of Turks from Greece. This of course increased the religious and linguistic homogeneity of the new Republic, but there was another consequence as well. Ottoman society had been multi-national and multi-religious, and there was a kind of specialization in functions and skills which followed communal lines. Greeks, Armenians, Jews had been in the forefront of economic enterprise. The greatly diminished presence of the Greeks, the Armenians, and similar groups dislocated the articulation of society and impoverished its texture. The new regime, moreover, in reaction against European business activity in the Empire—activity which they condemned as exploitation—tended to favour protectionism and autarchy. This tendency was accentuated following the world economic crisis which began in 1929, and when restrictions on the taxation of foreign imports imposed by the Lausanne Treaty expired. In 1931 Mustafa Kemal published a manifesto which set out the ideology of Kemalism, the tenets of which, as has been said, were incorporated in 1937 in Article 2 of the Constitution. One of these tenets was etatism. As Mustafa Kemal declared in words which were repeated in the programme of the Republic People's Party adopted in 1935,

Although considering private work and activity a basic idea, it is one of our main principles to interest the State actively in matters where the general and vital interests of the nation are in question, especially in the economic field, in order to lead the nation and the country to prosperity in as short a time as possible. The interest of the state in economic matters is to be an actual builder, as well as to encourage private enterprises, and also to regulate and control the work that is being done.

This was a very ambitious project. Under Kemal and his successors, the state established investment banks to finance and direct enterprises in the most important sectors of the economy. It also became itself an industrial entrepreneur in a big way. The results were not particularly brilliant, and by the time the Republican administration left office it had certainly not succeeded in discharging the responsibility it had, with such assurance, taken upon itself, of finding work for and ensuring a tolerable level of sustenance for the citizens whose numbers were so relentlessly increasing.

The Democrats, their successors, imagined escaping the economic stagnation they had inherited by spending large amounts of money in order to stimulate investment both in agriculture and in industry. It is true that output in both increased appreciably. However, the money which made all this activity possible came from budget deficits—in other words, the Government was printing money. In the decade of Democrat rule, it has been calculated that while national income grew by 200 per cent, the money supply increased by 408 per cent. Other statistics reinforce the picture of an economy afflicted with a severe, government-created inflation: the general index of wholesale prices increased from 46 to 126, and that of the cost of living in Istanbul from 54 to 133; foreign trade was also in deficit, exports amounting to some $320 million but imports to about $468 million.[5] The Democrats' economic strategy was based on the notion that governments had no need to observe the financial disciplines to which private persons are subject, that they could plan the path of a whole economy, and reach their appointed goal by manipulating the supply of money and credit. This was a new discovery, but once made, it had to be embraced by all who aspired to act on the political scene. Similarly, the voters for whose suffrages the politicians competed were taught to expect that these politicians would deliver work, sustenance, and increasing prosperity through these new magic arts. Menderes opened another Pandora's box. In order to sustain his support in the country-side, he paid high prices—much higher than those ruling on world markets—for agricultural produce. These policies created new expectations which were to change the whole tone of public life in the following decades. But it was not possible for the parties—certainly not for the Democrats—to deliver.

When the Democrats were driven out of office in 1960, the population had increased from the 21 million of 1950 to 27.7 million and the economy simply could not provide work for the steadily increasing multitudes. The situation, the Ankara correspondent of *The Times* wrote in 1962, was 'like a jungle of interrelated nightmares':

Of these perhaps the worst is unemployment; or to use the phrase now current, 'concealed underemployment'. This is not a matter of pockets here and there; the whole country is affected by it. It breeds a widespread paralysis and loss of power. All over Turkey one sees able-bodied men sitting in coffee-houses during working hours. Partly it is an atavistic inertia; partly it is the commercial stagnation which followed abandonment of the inflationary policies of Menderes, basically, it is the result of a population increasing too fast for the number of jobs to go round. 'Turkey', said a ragged individual mournfully to your Correspondent the other day in Ankara, 'is the workless country'.[6]

After 1950, two elections took place which the Democrats won by large majorities: in 1954 they won 505 seats against the Republicans' 31, in 1957 424 against 178. To judge by these figures one may conclude that Menderes and his policies were highly popular. One cannot however tell whether the figures are a simple, straightforward index of popularity. In the early years, before the second election, the impression of widespread prosperity produced by the Democrats' economic policy may have attracted suffrages. Again, the Democrats presented a much less hostile face than their predecessors to religion and men of religion. To the cause of religion great numbers were still attached, in spite of sustained Kemalist indoctrination, and of the difficulties officially put in the way of Islam. One month after assuming power, the Democrats allowed the call to prayer in Arabic, hitherto a criminal offence. The measure, it is interesting to note, was unanimously approved in the Grand National Assembly, the Republicans fearing that opposition to it would alienate from them large numbers of voters. The Democrats adopted other similar measures: the ban on the broadcasting of religious programmes was lifted, and the Koran was recited on the radio; religious education became virtually compulsory in schools; the budget of the Presidency of Religious Affairs constantly increased, and some 15,000 new mosques were built, under Democrat rule; Sufi brotherhoods, suspect to the Kemalists, now revived; and schools for the training of imams and preachers, as well as an Advanced Islamic Institute, were established. In following this policy the Democrats were acting with, and not against, the grain of society. This no doubt enhanced their popularity and led many to go on voting for them. Is it not likely, however, that the snowball effect here again operated, and that large numbers went on voting for the Democrats because they were the government, with all the power that a government in a country like Turkey was believed to have—until then at any rate—to bestow rewards on supporters and punishment on opponents? The fact remains that when the Democrats were ousted, in 1960, it was not through a swing of the pendulum bringing about electoral defeat, but through a military *coup d'état*.

When they became ensconced in office the Democrats did not hesitate to use their power in order further to diminish their rivals. After 1950 the officer corps was purged. In 1953, when attacks on the Democrats' policies were becoming more vocal both in the Assembly and outside, the Government took draconian action. In December 1953 a law was passed to deprive the Republican People's Party of all moneys, movable and immovable properties in its possession, to compensate the state 'for past misappropriation of public funds'. The People's Houses were also confiscated. All this made it impossible for the Party to publish its newspaper or to function at all.

The Democrats' heavy-handed rule was in the tradition established by the Young Turks and by Mustafa Kemal. In this tradition, for a government to brook any opposition was a fatal sign of weakness. Similarly those who oppose the government—and of course aspire to supplant it—will go to whatever lengths they can in order to weaken and destroy it. Hence if overt opposition is at all allowed by the government, its style will be strident, unrestrained and violent. The supporters of the Republican People's Party, both in and out of the Assembly, attacked and vilified the Democrat administration and all its works. Hostilities between Democrats and Republicans mounted particularly after the 1957 elections when the Republicans increased their representation in the Assembly. There was resort to mob violence, as when Inönü was bodily attacked in May 1959, in the countryside, and in Istanbul. The Government used the army and the police in order to repress demonstrations by Republican sympathizers, including university students who had been politicized by their teachers. In April 1960 Democratic deputies introduced a bill setting up an investigation of the Republican People's Party and of the press. The Republicans vociferously opposed the bill and walked out of the Assembly. The Democrats then passed a law appointing an Investigation Committee, packed with Democrats, which was given powers to imprison anyone, to shut down any newspaper, and suspend any law, if the Committee considered that they interfered with its work. University students at Istanbul and Ankara demonstrated at the end of April. The police and the army were ordered to break up the demonstrations. Clashes ensued in which many were injured, and some killed. The Government ordered universities to be closed. Inönü made a declaration vaguely calling for the army to intervene in order to 'save democracy'. These events in turn led students and teachers at the War College to set in train a conspiracy to overthrow the government. The army commander, General Cemal Gürsel, put himself at the head of the conspiracy, and went on leave in order to devote himself fully to the organization of the plot. On 27 May, the plotters struck. A group of officers who commanded key units in Istanbul and Ankara, led by Gürsel and making use of the War College cadets, arrested the President of the Republic, the Prime Minister, some of his colleagues, and Democrat deputies.

Martial law was declared and the country submitted to the *coup d'état*.

Gürsel and thirty-eight army officers constituted themselves into a National Unity Committee which assumed full powers of government. The Committee carried out a purge of Democrat officials, army officers, and university teachers. Thus, 90 per cent of the generals and 40 per cent of the colonels and majors were retired. The Committee also abolished the Democratic Party and confiscated its funds. A High Court composed of both civilians and officers was established to try leading Democrat ministers, deputies, and officials who had been arrested at the time of the *coup d'état* and afterwards. Those indicted before the High Court numbered 592, and included Bayar, Menderes, and other leading ministers. The defendants were charged with subverting the Constitution, corruption, illegally expropriating private property, inciting riots against the Greeks of Istanbul during a crisis over Cyprus in 1955, inciting attacks against leaders of the Republican People's Party, and with other crimes. Fifteen of the accused were sentenced to death, and the sentence was carried out in September 1961 on Menderes, Zorlu (the foreign minister), and Polatkan (the finance minister). The others, including Bayar, the ex-President of the Republic, had their sentences commuted to life imprisonment. Thirty-one others including ministers, members of the Investigation Committee whose activities had sparked off the *coup d'état*, and the former governor of Istanbul, were also sentenced to life imprisonment. Four hundred others were sentenced to lesser terms and 123 acquitted.

The *coup d'état* meant the abolition of the 1924 Constitution. To replace it, the National Unity Committee instituted a Constituent Assembly which debated a draft constitution which afterwards was submitted to, and approved by popular referendum in July 1961. This Constitution differed a great deal from the earlier Constitution of 1924. The academics and legal experts who produced the draft put before the Constituent Assembly were among the articulate opponents of the fallen regime which, in their view, had been profoundly reactionary and autocratic, and had sought to undermine Kemalism in various ways. Their draft constitution was on the one hand designed to prevent the rise of such an autocratic regime, and on the other to articulate what might be called a neo-Kemalism which gave the state much greater powers to affect— of course, it was hoped, for the better—social and economic life. In this, the new Constitution mirrored the quasi-socialist views of the junior army officers who were active in overthrowing the regime and who proliferated in the National Unity Committee.

The keynote of the new Constitution is heard in the preamble, the third paragraph of which states that it was

Guided by the desire to establish a democratic rule of law based on juridical and social foundations, which will ensure and guarantee human rights and liberties, national solidarity, social justice, and the welfare and prosperity of the individual and society.

These very ambitious and far-reaching objectives are spelled out in the body of the Constitution. Article 2 declares:

The Turkish Republic is a national, democratic, secular, and social State governed by the rule of law, based on human rights and the fundamental tenets set forth in the preamble.

To realize how much more extensive were the aims of the 1961 Constitution when compared with its predecessor, it is enough to compare this Article 2 with Article 2 of the earlier Constitution. This Article, following an amendment of 1928 which removed the stipulation that Islam was the religion of Turkey, read: 'The Turkish State is republican, nationalist, populist, etatist, secular, and reformist. Its official language is Turkish and its capital is the city of Ankara.' The earlier Constitution thus does not describe the state as 'social' or refer to 'human rights', and it does not have a preamble which aspires to establish 'social justice' and secure 'the welfare and prosperity of the individual and society'.

These large aspirations, laying great burdens on the fiscal and administrative capacity of the state, are unambiguously spelled out in great detail. Thus Article 10:

Every individual is entitled, in virtue of his existence as a human being, to fundamental rights and freedoms, which cannot be usurped, transferred, or relinquished.

The State should remove all political, economic, and social obstacles that restrict the fundamental rights and freedoms of the individual in such a way as to be irreconcilable with the principles embodied in the rule of law, individual well-being, and social justice. The State prepares the conditions required for the development of the individual's material and spiritual existence.

The State, Article 37 lays down, 'shall adopt the measures needed to achieve the efficient utilization of land and to provide land for those farmers who have no land, or own insufficient land'. The State, furthermore, 'shall assist farmers in the acquisition of agricultural implements'. Article 39 makes provision for nationalization of those private enterprises 'which bear the characteristics of a public service'. Article 41 orders that economic and social life 'shall be regulated in a manner consistent with justice, and the principle of full employment', so that everyone shall enjoy 'a standard of living befitting human dignity', and Article 42 further emphasizes that the State shall 'adopt measures to prevent unemployment'. Other rights and entitlements against the State are also specified. Article 48:

Every individual is entitled to social security. The State is charged with the duty of establishing . . . social insurance and social welfare organizations.

Article 49:

It is the responsibility of the State to ensure that everyone leads a healthy life both physically and mentally, and receives medical attention.

Article 50:

One of the foremost duties of the State is to provide for the educational needs of the people.

Finally, the whole of this imposing and grandiose erection is crowned by what is clearly meant to be the keystone of the whole edifice. Article 129:

Economic, social and cultural development is based on a plan. Development is carried out according to this plan.

Provision is made in the same Article for a State Planning Organization which is to prepare and execute, to apply and revise, and to take measures designed 'to prevent changes tending to impair the unity of the plan'.

This very ambitious, not to say visionary, *dirigisme* would, if applied in earnest, take centralization, already so prominent in the *tanzimat*, under the Young Turks and Atatürk, to new undreamt-of heights. The authors of these social and economic provisions of the Constitution no doubt believed that they were simply extending the scope of etatism and populism, two of the principles of Kemalism, which the Democrats, so they believed, had emasculated. As described by Mustafa Kemal in his manifesto of April 1931 which set forth the 'fundamental and unchanging principles' of Kemalism, etatism seeks

to interest the State actively in matters where the general and vital interests of the nation are in question, especially in the economic field, in order to lead the nation and the country to prosperity in as short a time as possible.

This definition of etatism figured in the Programme of the Republican People's Party. Populism, on the other hand, though listed in the Programme (along with State Socialism) as one of the defining characteristics of the Party, was not given a similar authoritative description. But those who drafted the Constitution could appeal to some early speeches by Mustafa Kemal in which he attacked 'the capitalism that seeks to swallow our very nationhood', spoke highly of 'the masses', and declared that 'the voice of the people is the voice of God'.[7] However, speeches and programmatic declarations apart, between the foundation of the Republic and 1961, there had been no attempt to legislate for a full-blown welfare state providing social-security benefits and medical care for every-one, let alone to impose on the Government the constitutional duty of ensuring full employment. Particularly during the 1930s there had been an attempt at central economic planning, and industries and foreign trade had come pretty well under detailed government regulation. But these policies and ventures, far from bringing industrial prosperity, had put the economy in a bureaucratic strait-jacket, and hardly justified the faith in planning and nationalization so evident in the new Constitution.

The neo-Kemalist provisions of the Constitution designed to undo what were believed to be reactionary Democrat policies were, however, incompatible with other features meant to make impossible the autocracy for which his opponents harshly blamed Menderes—but which Menderes had himself inherited from the Republicans. Thus the Constitution, in Article 120, declared that universities were to enjoy academic and administrative autonomy. Given the politicization of university teachers and students, it was not to be expected that governments would long forbear from intervening in university affairs. Similarly, Article 121 declared broadcasting and television stations to be 'autonomous public corporate bodies', presumably in order to prevent a government from using them as a propaganda tool. Given, however, the intemperateness and violence of the party politics as manifested in the Menderes decade, and that the intellectual classes were among the most prominent party-political warriors, how sensible or realistic was it for this Article to enjoin that broadcasts 'shall be made along the principles of impartiality'? How would such a provision be enforced? Again, how likely was it that governments would remain indifferent for long to politically adverse broadcasts, when such indifference would be widely construed as weakness, and thus *ipso facto* diminish their prestige? The most important means by which the National Unity Committee and their advisers thought to prevent governments abusing power was a change in the electoral law which introduced a form of proportional representation: each political party would receive seats in proportion to the votes cast in favour of its candidates, province by province. This arrangement naturally facilitated the rise of small parties, and made the attainment of a parliamentary majority by a single party, which had been the norm since the foundation of the Republic, much more difficult. The new constitutional scheme was thus inherently self-contradictory. On the one hand it aspired to a planned welfare state, and on the other it introduced electoral arrangements which made strong centralized planning impossible.

All of these innovations, then, would make Turkey more difficult to govern, and more subject to political instability. A harbinger of things to come was that in the referendum over the constitution the majority in favour was paper-thin: 6,348,191 voting for it, with 3,934,370 against, and 2,412,840 abstaining. It was not that voters necessarily had firm views about the welfare state, the State Planning Organization, or university autonomy. Rather, we may think those who abstained or who voted against were registering disapproval or dismay that a government which had seemed to pay some attention to their interests and preferences had been forcibly toppled. It has been said that Menderes and his two ministers were sent to their deaths in September 1961 in consequence of the referendum results in the previous July. Be this as it may, the fact remains that Menderes had released the Keynesian genie from its bottle. Dormant rural and provincial electorates now acquired an unappeasable appetite for official subventions. Thereafter politicians, when unconstrained by military intervention or the threat thereof, would not be able to resist the temptation to print money as a way of currying favour with voters solicited by their rivals, or as a means of escaping from economic difficulties. Thus as a result of Democrat policies Turkish politics were becoming volatile. A material cause of this volatility lay in the relentless increase in the population, which from 27.7 million in 1960 became 48 million in 1983—an increase which the Turkish economy had no capacity to cope with. An even greater cause of volatility was the increase in urban population: that of Istanbul, for instance, rose from 800,000 in 1940 to over five million in the mid-1980s. *Pari passu* with this increase went an equally relentless rise in the number of university students: for example from 65,000 in 1960 to 168,000 in 1970. These students were the products of the Kemalist school system where the official ideology of the regime was systematically inculcated by teachers who were themselves the product of the state teachers-training colleges.

Schoolboys learned the virtues of etatism, nationalism, and revolutionism—all components of Kemalism, listed in the Constitution, and expounded in the official programme of the Party. In universities which were modelled on those of continental Europe, with their absence of individual tuition and tutorial care, the milling multitudes of these students, attracted by activist ideologies whether of the Left or the Right, often spread by their own professors, would spearhead radical and revolutionary movements of various kinds, and engage in political violence inside and outside the campus. In brief, for the various reasons here set out, Turkish politics after 1961 would be very different from what they had been under Atatürk, Inönü, and Menderes. Writing in 1966, Dankwart Rustow shrewdly formulated the emerging problems:

What is now at stake is not just the restructuring of a limited political élite; rather, the issue now is no less than the admission to full political participation of the lower classes in the cities and of the peasant masses in the Anatolian villages.[8]

Some two decades and a half later, it is still doubtful whether these masses have been assimilated into Turkish politics in a manner such that they represent not a threat, but a buttress for the political order.

The two decades or so which followed the promulgation of the new constitution saw periodical army intervention, a succession of unstable coalition governments, high inflation with all the economic disorder, and the widespread discontent which this inevitably brings, and increasing political violence fomented by various terrorist factions, in many cases drawing inspiration and sustenance from abroad. The National Unity Committee, which had assumed power after the *coup d'état* of 27 May 1960, was by no means united. Some of its junior members favoured radical policies to be carried out by the armed forces; the generals on the Committee, however, favoured the re-establishment of civilian rule. The generals feared a new *coup d'état* by the radicals in November, and they succeeded in purging the Committee of fourteen suspected members who were arrested or put under house arrest. But this action by no means put an end to dissension among the officers. Another incident betokening division among the officers took place in June 1961. An Armed Forces Union had been formed shortly before, both to control dissident officers and to keep an eye on the National Unity Committee. One of the leaders of the Armed Forces Union, commander of the air force and a lieutenant-general, was suddenly ordered by the Committee to relinquish his command and go to Washington to head the Military Mission. The Union presented Gürsel, who was President of the Committee, Head of State, Prime Minister and Commander-in-Chief, with an ultimatum. Gürsel complied, the air force commander was reinstated, some members of the Committee dismissed from their army commands, the Minister of Defence, a general, was removed from the cabinet, and a number of generals retired. The Chief of Staff, General Sunay, also began participating in the Committee meetings. These actions did not, however, end dissension within the armed forces. Commanders were aware that junior officers entertained radical ideas and were ready to act on them. To avoid this threat, which became apparent after the general elections in October 1961, party leaders were forced to sign a statement agreeing not to seek an amnesty for the Democrat leaders who had been tried and sentenced, not to reinstate the members of the National Unity Committee who had been retired the previous November, to elect Gürsel as President of the Republic, and to accept Inönü (whose party had the largest number of seats, but not a majority) as Prime Minister. Some few months later, in February 1962, another coup was mounted by a colonel who assembled a force composed of cadets from the War and Gendarmerie Colleges and some armoured units in Ankara. The coup failed. The same colonel mounted another coup in May 1963. This coup also failed, and together with his principal collaborators he was tried and executed.

All the while, however, the civilian governments which issued from the October 1961 elections were under pressure from the armed forces and obliged to attend to their behests. Also, the last coup attempt occasioned the proclamation of martial law in Ankara, Istanbul, and Izmir; it was maintained until July 1964. Fortified no doubt by the existence of martial law, Sunay sent, on 12 November 1963, a letter to the Chairman of the National Assembly complaining that certain political parties were attacking the 27 May 'revolution', the army, and its commanders, in an attempt to drive a wedge between it and the country. The soldiers were unhappy, and should this state of affairs persist there would be a danger of armed revolution. The General demanded an answer by 22 November. Promptly, by the deadline set, the party leaders published a declaration swearing fidelity to the 'revolution' and promising to do nothing to provoke army intervention.

As has been seen, the 1961 electoral law introduced proportional representation in elections for the National Assembly. As a result, parties began increasing in number. The two parties of the 1950s subsisted during the 1960s and 1970s. The Democratic Party had been, of course, destroyed by the *coup d'état* and the trials which followed, but a new party appeared very soon in February 1961. The Justice Party became the principal heir of the Democrats and the legatee of much of its electoral support. The Republican People's Party continued to be led by Inönü until 1972. Other parties emerged in the October 1961 elections: the New Turkey Party which also aspired to succeed the Democrats and which polled 13.7 per cent of the vote, obtaining 65 seats; and the Republican Peasants Nation Party which, in an earlier *avatar* during the 1950s had been accused of subversion in the guise of religion and which, obtaining 14 per cent of the vote, gained 54 seats. As for the two major parties, the Republicans gained 173 seats with 36.7 per cent of the vote, and the Justice Party 158 seats with 34.8 per cent of the vote. It was clear that no single party had a majority. The army insisted that Inönü should be Prime Minister and he formed a government in coalition with the Justice Party, then led by a retired general, Ragip Gümüşpala. In June 1962 the coalition failed: Inönü put together another coalition with the New Turkey Party, the Republican Peasants Nation Party, and some Independents. This coalition collapsed in November 1963, and Inönü formed a third coalition with some Independents. In February 1965 this coalition was defeated on a vote for the budget, and Inönü resigned. This ministry was the last in which he was to hold office.

In the meantime, the leader of the Justice Party, Gümüşpala, had died in June 1964. He was succeeded by Suleiman Demirel, an engineer. Demirel was considered a moderate, who would not arouse the suspicion that the Justice Party was opposed to the secularism espoused by the army, that it was intent on policies favouring religion, or would seek revenge for what had been done to Menderes and his colleagues. The President, however, designated an Independent Senator who had been elected on the Justice Party list, Suat Hayri Ürgüplu, to form a government. It was again a coalition, the fourth since 1961. In this coalition, the Justice party was predominant, and its head served as Deputy Prime Minister. Other ministers were drawn from the Independents, the New Turkey Party, the Republican Peasants Nationalist Party, and the National Party, a small group which had seceded in 1962 from the Republican Peasants Nationalist Party.

In the elections which took place in October 1965, the Justice Party obtained 52.9 per cent of the vote and 240 seats. The Republicans' vote slumped to 28.7 per cent which gave them 134 seats. Demirel became Prime Minister of a Government all drawn from the Justice Party. Coalition governments between 1961 and 1965 had been short-lived and unable to formulate and follow coherent policies. This incoherence and instability might be expected to vanish when a single party commanded a safe majority in the Assembly. But even though Demirel remained Prime Minister until March 1971—having won another general election in October 1969—Turkish politics showed increasing signs of factionalism and polarization, erupting in street violence in which various extremist groups clashed with one another. When he became leader of the Justice Party in 1964 Demirel, as has been said, was regarded as a moderate. As understood by the officers, this label meant that he would not seek to pursue anti-secularist policies. But Demirel's moderate stance was ambiguous. In order to defend himself against accusations of freemasonry—which for the partisans of Islam carried echoes of Young Turk and Kemalist irreligion—

Demirel would say that he had been born into a family 'that does not sit down to breakfast before reading the Holy Koran'. In the October 1965 elections the Justice Party exploited the theme of Islam, in order to appeal to the same voters in the countryside and provincial towns who had supported the Democrats. The fact that Demirel had been born in a village called Islamköy (Village of Islam) and that his father had performed the pilgrimage to Mecca was made much use of, in the same way as the Democrats had used the fact that Menderes escaped unscathed from an air crash in 1959, in which many died, to claim that he had been elected Prime Minister by God. The exploitation of the religious theme by the Justice Party was all the more successful in that the Republicans had a few months earlier decided to campaign with the slogan that they were 'left of centre'. This enabled the Justice Party to tar the Republicans with the Communist brush. As their slogans claimed: 'Left of centre is the road to Moscow', and 'We are right of Centre and on the path to God'. Again, in a speech at the end of June 1965, Demirel declared: 'Communism will not enter Turkey because our population is 98 per cent Muslim. We must be able to call ourselves a Muslim nation.' As Menderes had discovered earlier, and as Demirel's successors were to establish later, discreetly playing the Islamic card could not but be a profitable gambit. In spite of years of indoctrination, Kemalism had not proved to be an adequate substitute for Islam, and had not succeeded in attracting the loyalty of the mass of voters.

The Republicans' 'left of centre' slogan was adopted under the influence of Bülent Ecevit, who had served as Minister of Labour in Inönü's 1961 administration. Ecevit's influence in the Party increased greatly following its poor showing in the 1965 elections. Ecevit considered that a new departure was absolutely necessary. The Party had for too long been identified with Kemalist doctrine and with the bureaucratic style of government which for so long obtained in Turkey. Ecevit believed that the Party had now to appeal to urban workers and to farmers, so as to be able to limit the influence of, and eventually defeat, the Workers' Party of Turkey. This Marxist party had been founded in 1961, and had succeeded in winning

3 per cent of the vote and 15 seats in the Assembly during the 1965 elections. Ecevit's strategy won out in the Party debates which followed these elections. In 1966 he was elected Secretary-General, and in 1972 he ousted İnönü (who died two years later) in the chairmanship of the Party.

The new-style leftist-leaning Republican People's Party, vying with the Marxist Workers' Party for the suffrages of the poor, contributed a great deal to the polarization and volatility of Turkish politics after 1965. So did the strategy of the Justice Party which tried to destroy the Republicans' electoral appeal by portraying them as Leftists in disguise who were bent on destroying the Islamic tradition of Turkey, and who were engaged in a dangerous flirtation with Communism, the ideology of Soviet Russia—a power which threatened the safety and the interest of Turkey.

Demirel's economic policy, harking back to that which Menderes had favoured a decade earlier, was to stimulate the economy, and thus retain the favour of the electorate, by means of deficit finance and foreign loans, and as happened under Menderes, the result was inflation and eventually devaluation of the Lira, with the attendant discontents and social stress to which inflation inevitably leads. These discontents had now become particularly dangerous. Migration from the countryside constantly and relentlessly increased the urban population. Housing, municipal, and social services were utterly unable to cope with the needs of the new migrants who were either unemployed, or earned a mere pittance, and who were compelled to huddle together in miserable shanty towns on the outskirts. Their presence and their natural susceptibility to leftist rhetoric served to make them into a volatile mass providing recruits to the subversive and terrorist movements which began to flourish in the second half of the 1960s. The spearhead of these movements was constituted by leftist and Marxist intellectuals and university students whose hopes for a revolutionary upheaval leading to the overthrow of capitalism and reaction received a great boost from the militant student movements which were such a prominent feature of the period in France, West Germany, Italy, Britain, and the United States.

In the late 1960s and early 1970s Turkey gave the impression of being in the throes of serious disorder. In 1968 student revolts erupted on various campuses. Agitation against Turkey's connection with the USA and NATO was naturally prominent. There was a demonstration in 1968 against the US Sixth Fleet, then on a visit to Istanbul, in which a student was killed. At the Middle East Technical University, students burned the car of the US Ambassador in the following year. On 16 February of the same year another demonstration took place against the Sixth Fleet in one of the principal Istanbul squares. The demonstrators, shouting 'Independent Turkey!', clashed with counter-demonstrators shouting 'Muslim Turkey!' The incident led to two killed and about two hundred wounded, and came to be known as Bloody Sunday. This clash indicates that street agitation was fomented not only by left-wingers, but also by equally organized Islamic or nationalist opponents. As was to become increasingly manifest in the 1970s, political disorder ceased to be simply a matter of street demonstrations leading at times to violent clashes. Rather, various groups, both left and right, began training armed commando groups to carry out terrorist actions, such as kidnapping and murder.

The conjugated effects of inflation, street disorder, and terrorism led to yet another intervention by the army. The 1961 Constitution had made provision for a National Security Council on which sat the Chief of the General Staff, commanders of the army, navy, air force, and gendarmerie, the Prime Minister, the Ministers of Defence, the Interior, and Foreign Affairs, under the chairmanship of the President. The existence of such a body, able constitutionally to express views on public affairs, gave a kind of legitimacy to intervention by the armed forces in politics. On 12 March 1971 the force commanders fearing, it is said, a possible left-wing coup to be carried out by junior officers, forced Demirel to resign. They simultaneously published an ultimatum signed by all four commanders and the Chief of the General Staff setting out the reasons for their move and what action political leaders had now to take. They said:

1. The Parliament and the Government, through their sustained policies, views and actions, have driven our country into anarchy, fratricidal strife, and social and economic unrest. They have caused the public to lose all hope of rising to the level of contemporary civilization which was set for us by Atatürk as a goal, and have failed to realize the reforms stipulated by the Constitution. The future of the Turkish Republic is therefore seriously threatened.

2. The assessment by the Parliament, in a spirit above partisan considerations, of the solution needed to eliminate the concern and the disillusionment of the Turkish Armed Forces . . . and the formation, within the context of democratic principles, of a strong and credible government, which will neutralize the current anarchical situation . . . are considered essential.

3. Unless this is done quickly, the Turkish Armed Forces are determined to take over the administration of the State in accordance with the powers vested in them by the laws to protect and preserve the Turkish Republic.

Please be informed.

Following this forcible intervention, the commanders of the armed forces proceeded to take other measures which would, in their estimation, rid the country of the evils their memorandum had described. As the second paragraph of the memorandum stated, they were looking to the people's representatives to take the steps necessary to remedy the situation. As in 1960, the commanders who directed the military intervention had no intention of taking on political and administrative responsibilities. It may have been different had more junior officers captured the leadership, as one may suspect from the successive attempts at a *coup d'état* in 1961–3. Now, too, the more radical elements in the armed forces had to be crushed. Five days after the ultimatum and Demirel's dismissal, five generals, one admiral, and thirty-five colonels were retired.

The force commanders were, however, determined to put right what they believed had gone wrong with the state and its institutions. One cause of the problem, they considered, was the 1961 constitution which resulted in weak, coalition governments, and which allowed free rein to organized subversion which sought to discredit the Kemalist legacy and destroy Turkish social and political institutions. The first requisite was thus to amend the Constitution in order to remove its most harmful provisions. Changes in thirty-five articles were therefore promoted and nine temporary clauses added. Their purpose was to strengthen the power of government to take action against threats to public order, to national unity, and national security; as well as to increase the autonomy of the armed forces and their freedom of action. The changes increased the powers to prohibit the exploitation of class, sect, religion, race, or language in order to promote divisiveness; they limited the right to form unions and associations; they gave the cabinet power, when authorized by the Grand National Assembly, to issue decrees having the force of law; they extended the length of time during which martial law could remain current without the necessity of parliamentary approval; they strengthened state control over the electronic media; they redefined the autonomy of universities so as to make it illegal for them to shield criminals; and finally they gave the National Security Council the right not only to present its views to the cabinet, but also to make recommendations.

The armed forces' ultimatum and Demirel's departure did not mean the end of public disturbances. Indeed, April saw new attacks organized by so-called urban guerillas. A highlight in these attacks was the kidnapping and murder of the Israeli Consul General in Istanbul in the second half of May. Martial law was proclaimed in eleven provinces at the end of April and remained in force for two years. Leftist groups and publications were banned and large numbers of suspected Leftists taken into custody, while the Marxist Workers' party was dissolved by court order in July 1971. Shortly before this, another party, the National Order Party, formed in 1970 by Professor Necmettin Erbakan, had also been dissolved on the score that its Islamist programme and activities violated the provisions of the Constitution, which prohibited advocacy or attempts designed to change the secular character of the state.

The authors of the 12 March ultimatum had left in being the Grand National Assembly with its majority drawn from the Justice Party. It was, however, clear that the soldiers wanted neither Demirel nor an administration drawn wholly from the majority party. Sunay, now the President of the Republic, asked Nihat Erim to form a new government. Erim was a member of the Republican People's Party from which he now resigned so as to head a reform, 'above-party' cabinet which would have the support of the armed forces. It was a coalition drawn from the Justice Party, from the Republican People's Party, and from the National Reliance Party which had been formed by seceders from the Republican People's Party when, under Ecevit's inspiration, it adopted a left-wing programme. The majority of Erim's administration was, however, non-party and drawn from the ranks of 'technocrats' and high-ranking officials. Little coherence was to be expected from such a cabinet which, anyway, found itself hemmed in on one side by an Assembly in which Demirel and his followers predominated, and by the soldiers on the other. This administration fell in December 1971. Erim formed another coalition drawn from the same elements as the earlier one. When it fell in April 1972, Ferit Melen of the National Reliance Party was asked to head a new coalition where, again, there were ministers drawn from the two main parties in the Assembly and from the Prime Minister's own party, as well as non-party ministers. This administration, suffering from the same incoherence and subject to the same constraints as its two predecessors, fell in April 1973. Its successor was headed by Naim Talu, a non-party man who had served as Minister of Trade in the outgoing cabinet. It was composed solely of Justice Party and non-party men, and its principal task was to carry on government until the general election which fell due in October 1973. If the soldiers had hoped that their ultimatum would lead to reform of the political and administrative institutions, to the eradication of inflation, and the enhancement of social stability, then the record of two and a half years in which four coalition governments succeeded one another to no discernible purpose must have grievously disappointed them. What kept the peace in the country were martial law and the amendments of the constitution on which the commanders had insisted; and it was their presence in the background, and fear of their intervention, which provided a restraint on the parties in the Assembly and kept them from pursuing their quarrels to a bitter end.

The October 1973 elections resulted in 185 seats for the Republican People's Party and 149 for the Justice Party. But two new parties now also gained a substantial number of seats: the National Salvation Party, founded in 1972, fervently Islamist and successor to the National Order Party dissolved the previous year, gained 49 seats; and the Democrat Party (to be distinguished from the Democratic Party which the *coup d'état* of 1960 had abolished) which was founded in 1970 by seceders from the Justice Party who disagreed with Demirel's policies, gained 45 seats. Another coalition was thus inescapable. Negotiations were complicated and long-drawn-out. Finally, in January 1974, Ecevit and Erbakan, leader of the Salvationists, agreed to form a government. The alliance was incongruous in that the Republicans had been, throughout their history, insisting on the Kemalist principle that Turkey was a secular state, while the Salvationists believed that only if the country went back to its Islamic roots would it become happy and prosperous. The two leaders, however, cobbled together a joint programme, wide in its scope and grandiose in its ambitions—a programme which, given the resources of Turkey and the capacities of its political leaders and its officials, there was not the slightest prospect of realizing. The tone of the programme, with its inflated and empty rhetoric, is caught in this passage:

The two parties which form the government know that it will not be sufficient merely to broaden the freedoms in order to strengthen democracy and society. We believe that a regime of social justice covering the whole of society is as necessary as political rights and liberties and the freedom of thought and belief. We know that extensive social justice rooted in liberty cannot be considered in isolation from rapid and balanced economic development. That is why we are determined to realize and secure democratic freedoms in the broadest sense simultaneously with extensive social justice and rapid economic development.[9]

Unfortunately the government which was to accomplish all these things fell after only nine months in office. It could not be denied however that it had accomplishments to its credit:

In March the government permitted the opium poppy to be sown in six provinces, despite threats from the US government that, as a consequence, it might cut off aid to Turkey. Under pressure from Washington, the Erim Government had prohibited poppy cultivation in 1972. This decision had been very unpopular in the country and became the symbol of Turkey's subservience to America. Its reversal by Ecevit was most popular for it was seen as restoring Turkey's dignity and independence, and it added to Ecevit's personal standing and prestige. Such was the sense of freedom and optimism that Mehmet Ali Aybar, the ex-chairman of the dissolved Workers' Party of Turkey, suggested that in view of Premier Ecevit's promise to establish a democratic society equal to any in Europe, the time was ripe to establish a new socialist party in Turkey.[10]

It became fairly clear soon after the coalition was formed that the two parties were suspicious of one another and that co-operation was out of the question. In July 1974 events in Cyprus took a turn which came to exert a direct influence on the fortunes of the coalition. A *coup d'état* of 15 July 1974, supported by the Greek military junta which had ruled in Athens since 1967, had overthrown the Cypriot President Makarios and put in his place another Cypriot nationalist even more in favour of *enosis* with Greece than Makarios. Turkey, a guarantor of the Treaty of Zürich of 1960 which had established a bi-national Cypriot republic, considered that the interests of the Turkish Cypriots, of which it was the guardian, were gravely threatened. Ecevit called upon the other guarantor, Britain, to join it in taking action against the subversion of the Treaty by the third guarantor, Greece. However, the British Labour Government, professing great faith in the emollient and healing powers of diplomacy, absolutely refused to join Turkey in military action to undo the effects of the *coup d'état*. Ecevit, strong in Turkey's rights as a guarantor state, sent Turkish troops to the island. They eventually occupied 40 per cent of the territory, and to this day, in the absence of a settlement, have remained ensconced there. This action raised Ecevit's popularity to great heights. He resolved to take advantage of this by resigning on 18 September, thus hoping to form a coalition which would prepare new elections, in which he would reap the fruits of his great sudden popularity.

Ecevit's rivals, however, had no intention of helping him in this manner. Again, long-drawn-out negotiations ensued, but with no result. Eventually, on 12 November, the President asked a professor of medicine, Sadi Irmak, to form an administration which would be 'above party'. Irmak did so, but could muster only seventeen votes in the Assembly when he asked for a vote of confidence. He naturally resigned, but he and his colleagues had to remain in office, as the main party leaders could not agree on forming a government. Eventually, at the end of March 1975, Demirel formed a coalition in which the largest contingent (sixteen) was furnished by his own party. The National Salvation Party contributed eight, the Reliance Party four and the Nationalist Action Party two. This last party had only three members in the Assembly and was led by Alparslan Turkeş. Colonel Turkeş had been prominent in the *coup d'état* of May 1960, and was one of the most radical among the junior officers who organized it. Shortly afterwards he was dismissed from the National Unity Committee and went into exile. He came back in 1963 and was involved in an abortive coup. In 1966 he became leader of the Republican Peasants Nation Party, the name of which was changed in 1969 to Nationalist Action Party. Turkeş, as has been said, inclined to radical causes with which his party became identified. Under his aegis and inspiration nationalist commando groups began to be formed and trained in order to oppose similar armed revolutionary groups on the left. Just as Erbakan and his Salvationists were incongruous partners for Ecevit, so Turkeş and his strong-arm group should have had no place in a government—even a Nationalist Front one, as Demirel called it—the other members of which eschewed the use of violence in politics. This kind of short-term opportunism practised by all parties made parliamentary and democratic politics an affair of *combinaziones, ad hoc* and *pro tem*, in which it is difficult to discern any principle except that of holding power for its own sake, and at any price. A case in point is the programme of the Nationalist Front Government. Ecevit had proposed to lower the voting age to 18: Demirel promised likewise. The Republican People's Party promised health and unemployment benefits for all workers: so now did the Nationalist Front.

But perhaps the most important vote-catcher in any future elections was the promise to reduce the price of artificial fertilizers. The price had risen sharply after the increase of oil prices in 1973 . . . The rich landowners . . . were forced to pay more, and they were unhappy with the situation. It was their votes—and the votes of peasants dependent on them—that could be won by Demirel's concession. Moreover Demirel did not wait long to fulfil his promise. On 19 April the Government announced that the prices of various types of fertilizer had been reduced by between 43 and 154 *kuruş* a kilo. By reducing the price, said Demirel, the Government had subsidized the landlords with about TL 5 billion.[11]

The Nationalist Front coalition lasted until the general elections in June 1977. During this period, inflation was rampant, the economy was in increasing difficulties owing to the scarcity of foreign exchange and the high price of oil which OPEC had engineered in 1973–4, and public order was beginning to be threatened by the organized gangs whose activities were shortly afterwards to rise to a terrifying crescendo. The partners in the coalition were able to do very little about such problems. They, however, became busy bestowing on their supporters posts in those departments over which they had ministerial control. The civil service—and what is more serious, the police—became highly politicized, to the detriment of good government and public order.

The June 1977 elections were a set-back for Demirel, producing 213 seats for the Republican People's Party and 189 for the Justice Party, while the National Salvation Party gained twenty-four seats and the Nationalist Action Party sixteen. Ecevit, having the largest number of supporters, formed a government, but not having an absolute majority, he soon fell. Demirel came back with his coalition, and they ruled for six months. But at the end of six months, some of his supporters became disgruntled by the prominent position in the government of the Salvationists, who had eight ministers, and by Demirel's reliance on the Nationalist Action Party, which, with all its thuggish predilections, now disposed of five ministeries. Enough of them defected to Ecevit who, with their help (obtained by giving them ten newly-created ministries), now formed another government which depended on a thin majority and which lasted until October 1979. The new administration faced the same problems of violence and economic stringency which in the past had afflicted both Ecevit and Demirel. In mid-term elections to the Senate, the Justice Party gained 30 seats out of 50, and five by-elections to the Assembly were lost by Ecevit. His majority in the Assembly disappeared, and he fell. Demirel succeeded him, forming a minority government in coalition with Independents, but still having to rely on the support of the Salvationists and the Nationalist Action Party. This government lasted until 12 September 1980, when a military *coup d'état*, the third in three decades, brought it to an end.

While, in the seven years between the elections of 1973 and the *coup d'état*, the politicians were playing musical chairs and keeping their network of clients in good repair, and possibly also extending and improving it, the country was experiencing increasing misery, stress, and fear. Short-term, high-interest foreign loans, the printing of money, and other specimens of financial mismanagement, occasioned by the desire to attract or reward supporters, or by mistaken confidence in the virtues of economic planning, allowed the economy to get out of control. The upshot was devaluation of the currency, steep and continuous increases in the price of foodstuffs and other necessities, and a mounting scarcity of basic imports—foreign currency was simply not available to pay for them. The increase in wholesale price levels between 1977 and 1980 is sufficient indication of the cumulative worsening of conditions:

1977	24%
1978	53%
1979	64%
1980	107%

After the *coup d'état*, in 1982, the General Secretariat of the National Security Council—whose military members organized and directed the overthrow of Demirel's government—published a substantial volume, *12 September in Turkey: Before and After*, a large part of which was a collection of reports culled from newspapers and other publications, portraying the economic and security situation of Turkey during 1977–80. In 1979, for instance, the press was reporting that there was a shortage of diesel oil, kerosene, and gasoline, that fuel-oil stocks were exhausted, that medicines were becoming rare, that factories were closing down, that the lack of diesel oil was affecting agriculture, and that the lack of fuel oil meant reductions in public transport which in turn led to the paralysis of urban life. The use of electricity was severely restricted, and the compilers humorously remarked that conferences on heavy industry—no doubt attended by important officials—were held by the light of kerosene lamps and candles.

Economic stringency was accompanied by a mounting number of terrorist attacks and murders, and violent riots, mounted by various left- and right-wing groups. The financing, arming, and organization of these groups was never fully clear, but there was a great deal of evidence that both funds and arms originated from sympathizers in Turkey and from interested foreign powers, Communist or Islamic. Both attacks and riots had been increasing before the ultimatum of 12 March 1971, and they resumed after the October 1973 general elections. It has been calculated that deaths from political violence in the years before the 1980 *coup d'état* were approximately as follows:

1975	35
1976	90
1977	260
1978	800–1,000
1979	1,500
1980	3,500[12]

The volume, *12 September*, mentioned above, comprises a grisly calendar, compiled from press reports, of the incidents which led to these violent deaths. A few highlights from this calendar include a mass May Day 1977 demonstration in

Istanbul organized by the Marxist Revolutionary Confederation of Trade Unions which led to armed clashes and a stampede in which thirty-four people lost their lives, shops being looted, and cars burned in the neighbourhood of the demonstration. Another armed clash, which led to the declaration of martial law in thirteen provinces, occurred in Kahramanmaraş in south-eastern Turkey on 22–5 December 1978. The clash involved Sunnis and Shiites—Sunnis favouring the right and Shiites the left. The clashes resulted in 109 dead, 176 seriously wounded, and 500 houses and shops destroyed. Again, in December 1979, four American servicemen were murdered on a beach by a 'Marxist-Leninist Armed Propaganda Unit'. In June 1980, the district chairman of the Republican People's Party at Nevşehir was assassinated; Ecevit and other party leaders went to the funeral and were attacked with stones and bullets; there were many wounded and panic ensued. In July, a bomb exploded near a mosque at Çorum, north-east of Ankara, during the Friday prayers, and the area of the mosque was sprayed with bullets. By evening 100 dwellings were ablaze and a progrom was in train against the Shiite supporters of the Republican People's Party who were suspected by the Sunni supporters of the National Salvation Party of perpetrating the morning's outrage.

Martial law was clearly unable to cope with the mounting disorder, and with the challenge to the Turkish constitutional order as laid down by Atatürk. The leaders of the armed forces considered themselves the guardians of this order, invoking Article 35 of the Turkish armed forces code which stated that their duty was 'to protect and safeguard the Turkish land and the Turkish Republic as stipulated by the Constitution'. The Chief of Staff, General Kenan Evren, in a speech on 12 September 1980, contrasted the casualties occasioned by terrorist attacks during the two previous years with those sustained during the crucial battle of Sakarya when the Greek army was routed in August 1921. The terrorism had resulted, he said, in 5,241 dead and 14,152 injured, while the battle had resulted in 5,713 dead and 18,480 wounded. It was not only the casualties which aroused the disquiet of the generals: even more serious were the unrestrained and open ideological warfare between leftist and Islamist groups, and their unbridled attempts to mock and bring into contempt the Kemalist legacy, and to herald civil war. Thus, in July 1980 the mayor of Fatsa on the Black Sea and his followers proclaimed the area a 'liberated' zone and purported to set up an autonomous commune. Again, barely a week before the military *coup d'état*, on 6 September, the National Salvation Party organized in Konya a Liberation of Jerusalem Day, when, according to the account given in *12 September*, during the demonstration

a big green flag (which symbolizes Islamic law) with Arabic writing was being paraded in front of the crowd. Those people with green and white skull-caps on their heads, prayer beads on their necks, and religious robes of a variety of colours on their backs were shouting all the way: 'Sheriat will come, savagery will go'. In the meantime, a group of youngsters, wrapped up to their eyes, were carrying green flags as well as placards featuring the Koran and pictures of automatic weapons for guerilla warfare. The mob was shouting various slogans such as 'Either Sheriat or death', 'Secularism is atheism', 'Islamic society, Islamic State', 'The Koran is the Constitution'.

Faced with these alarming disorders, political leaders, it seemed to the officers, simply engaged in endless futile bickering. The Assembly was a forum in which members traded insults rather than engaged in reasoned debate. When the term of the President of the Republic was expiring, it proved impossible to elect a successor, for no matter how many ballots took place the requisite majority for a candidate could not be obtained: a state of affairs emblematic of the deadlock and paralysis at the heart of government.

The military leaders—the Chief of Staff, Kenan Evren, and the four commanders of the armed forces who sat on the National Security Council—took over the government. General Evren became head of state and chairman of the National Security Council. The Demirel Government was dismissed, and the Prime Minister and his chief opponent, Ecevit, were briefly kept in custody at an army base some distance from Istanbul. Erbakan and Turkeş were also arrested. The National Security Council, shorn of its civilian members, became the highest political authority in the land. Martial law covered the whole country; in the civil service, which had been politicized, a very large number were dismissed; two trade-union federations regarded as propagators of extremism, were dissolved; strikes were banned; mayors and local councillors suspected of extremism were also dismissed. Political parties were closed down, and a year later abolished and their funds confiscated. In the four years following the *coup d'état*, 178,565 persons were arrested, 64,505 detained, 41,727 sentenced to various terms in prison, and 326 death sentences pronounced. The authorities confiscated 26 rocket launchers, 1 mortar, 638,000 hand guns, 4,000 automatic pistols, 48,000 rifles, 7,000 machine guns, and 6 million rounds of ammunition. Under such ministrations, Turkey began to enjoy a measure of peace and quiet.

The generals who organized the *coup d'état* of 12 September 1980, like their predecessors in 1960 and 1971, did not wish for a military regime to be established. They rightly realized that the armed forces did not have the capacity to manage the complex legal and administrative institutions of a large and populous country and, at the same time, assume responsibility for defence and security. As early as July 1981, they announced the establishment of a National Consultative Assembly whose main business was to draft a new constitution. The Assembly appointed a committee which presented a draft constitution in July 1982. The draft was considered, amended, and approved by the Assembly and the National Security Council—consisting of Evren and the four heads of the armed forces—and then submitted to a referendum in November 1982. Over 91 per cent of voters participated and of these over 91 per cent again approved the constitution. Voting in the referendum was compulsory, and failure to vote was punishable by a fine of 2500 Liras, and deprivation of voting rights for a period of five years. It was also forbidden to campaign against approval of the Constitution. Evren, also, was to the fore in campaigning for the Constitution. A provisional article in it stipulated that, on its adoption by the people, the then head of state, i.e. General Evren, would *ipso facto* become, for the seven years following, President of the Republic. This was an exceptional provision, since the new Constitution laid it down that a President of the Republic was to be elected by a two-thirds' majority of the Grand National Assembly. President Evren took office immediately following the referendum. In 1983 there followed a law on elections and one regulating the operations of political parties. In November 1983 the first elections under the new regime were held.

The new Constitution is a much longer and more detailed document than its predecessors. It shares some features with each of them, but its provisions clearly also attempt to remedy what its authors, and particularly the military leaders, believed to be the shortcomings of the 1961 Constitution—shortcomings which they considered to be responsible for the disorder of Turkish politics during its currency. One of the most prominent features of the new Constitution relates to the greatly enlarged powers of the President. He can return to the Assembly laws for its reconsideration; he can submit to a referendum legislation amending the constitution; he can appeal to the Constitutional Court for the amendment in whole or in part of laws or decrees which he considers unconstitutional; he can call new elections for the Assembly. The President decides on the use of the armed forces; he appoints the Chief of the General Staff on the Prime Minister's proposal; he calls meetings of the National Security Council and presides over them; he appoints members of the Higher Education Council and university rectors; he also appoints members of the Constitutional Court, the Chief Public Prosecutor, and his Deputy, members of the Military High Court of Appeal, of the Supreme Military Administrative Court, of the Supreme Council of Judges and Public Prosecutors, and one-quarter of the members of the Council of State. The Constitution also makes provision for a State Supervisory Council, whose members and Chairman are appointed by the President. This new body is attached to the President's office and seems to have wide powers. Upon the President's request it will investigate and inspect 'all public bodies and organizations', all enterprises whose capital is more than 50 per cent publicly owned, also 'public professional organizations, employers' associations, and labour unions at all levels, and public-benefit associations and foundations' (Articles 104 and 108).

The ability of the armed forces to have a say in the formulation of policies, already enhanced by the amendments to the 1961 Constitution introduced following the ultimatum of March 1971, is here entrenched and magnified. It is the President who usually presides over the meetings of the National Security Council. He also draws up its agenda, and in doing so he takes account not only of the Prime Minister's proposals, but also of those put forward by the Chief of the General Staff. The National Security Council submits to the Council of Ministers

its views on taking decisions and ensuring necessary co-ordination with regard to the formulation of the National Security policy of the State. The Council of Ministers shall give priority consideration to the decisions of the National Security Council concerning the measures that it deems necessary for the preservation of the existence and independence of the State, the integrity and indivisibility of the country, and the peace and security of society. (Article 118).

These provisions relating to the President and the National Security Council are at the heart of the new arrangements, which were no doubt designed to prevent a repetition of the state of affairs which brought Turkey to the brink of civil war between 1961 and 1980. But the Constitution includes other significant buttresses which, it must have been thought, would enhance political stability. Thus Article 28 made it an offence to publish anything which threatens the integrity of the state and its internal and external stability, and which tends to incite riot and insurrection (Article 28). Trade unions are forbidden to engage in political activity, or to receive support from, or to give support to, political parties (Articles 52). Politically motivated strikes and lock-outs are also forbidden (Article 54). Political parties are forbidden from supporting the domination of a class or group, from advocating any kind of dictatorship, and from adopting programmes contrary to 'the principles of the democratic and secular Republic' (Article 68). No one convicted 'of involvement in ideological or anarchic activities, and incitement and encouragement of such activities' can, if pardoned, be elected a deputy (Article 76). Local government, the responsibility of locally elected councils, is nevertheless subjected to the 'administrative trusteeship' of the central government (Article 127). As has been seen, the President appoints the members of the Higher Education Council; however, candidates to these posts are nominated not only by universities, but also by the Council of Ministers and by the Chief of the General Staff. The President, again, appoints university rectors. In addition, however, deans of faculty as well as all other teaching staff are appointed and removed by the Higher Education Council, itself a body wholly appointed by the President; the Council, indeed, plans, organizes, administers, and supervises education and research in universities (Articles 130 and 131). University teachers are forbidden to belong to a political party (Article 68). This is a far cry from the 1961 Constitution which simply declared that universities enjoyed academic and administrative autonomy and allowed teachers to belong to political parties. In that Constitution radio and television were similarly to enjoy administrative autonomy; the later Constitution removes this privilege from the public corporate body administering the media; it is charged, instead, with ensuring not only that broadcasts are impartial, but also that they will safeguard the independence and integrity of Turkey, the 'peace of society', public morals, and 'the fundamental characteristics of the Republic' as laid down in Article 2, namely that it is a 'democratic, secular, and social State' governed by the rule of law, respecting human rights, 'bearing in mind the concepts of public peace, material solidarity, and justice', and loyal to 'the nationalism of Atatürk' (Article 133).

It is the President, as has been seen, who now appoints all the members of the Constitutional Court, whereas under the earlier Constitution only two appointments were in his gift. This Court was created by the 1961 Constitution. The Court had power to examine the constitutionality of laws and by-laws, and thus of decisions reached under them, and to annul them if necessary, and its rulings were final. The manner in which the Court exercised its powers in the disturbed conditions of the 1970s aroused disquiet, particularly in the armed forces. The

Court ruled as constitutional a bill, which had been rejected by the Assembly in 1974, and which allowed the release of 4,000 political prisoners, who were in consequence released. The Court also ruled as unconstitutional the State Security Courts which had been established in 1973 with the purpose of dealing speedily with the cases of those accused of political offences, and they were dissolved in 1976. In the 1982 Constitution, the Court was forbidden from ruling on the constitutionality of, or substance of, decrees issued during a state of emergency or under martial law. It did not have the power to annul laws unless the President of the Republic, or the members in the Assembly of the party in power, or of the main opposition party, and a minimum of one-fifth of the Assembly members, applied to it for a ruling. Further, in annulling a law or a decree in whole or in part, the Court 'shall not act as a law-maker and pass judgment leading to new implementation' (Article 153).

The 1982 Constitution did not, however, cancel or abandon all of the main features of its predecessor. Article 2 of the 1961 Constitution declared that the Turkish Republic was a 'social' state. The same term occurs in Article 2 of the later Constitution. What this word signified for the authors of the 1961 Constitution is apparent from the many duties laid on the state to prevent unemployment, to provide social welfare, medical services, and housing, and to plan the economy to this end. Considerable traces of the same outlook subsist in the later document. Thus the state is charged with ensuring the payment of fair wages (Article 55), to meet the need for housing (Article 57), to provide social security (Article 60), to 'protect artistic activities and artists' (Article 64). There is, however, recognition that the fulfilment of these objectives, desirable as they might be, is limited by the available financial resources, 'taking into consideration the maintenance of economic stability'. Even so, here too the same *dirigiste* ambition which characterized Atatürk and his Party, and which was so prominent among those who overthrew Menderes, finds unmistakable—and surprising—expression in Article 23 which declares that freedom of residence may be restricted 'for the purpose of . . . promoting social and economic development'. Finally, Article 166 is categorical and comprehensive in ambit. It lays down:

The planning of economic, social, and cultural development, in particular the speedy, balanced, and harmonious development of industry and agriculture throughout the country, and the efficient use of national resources on the basis of detailed analysis and assessment, and the establishment of the necessary organization for this purpose, are the duty of the State.

The Kemalist outlook and legacy is also present, albeit not without ambiguity, in other features of the Constitution. In the preamble to the Constitution, which forms an integral part of it (Article 176), it is declared that one of its objects is to ensure that the Turkish Republic 'attains the standards of contemporary civilization, as a full and honourable member of the world family of nations'. Here is manifest the old aspiration of the Young Ottomans and the Young Turks to which Mustafa Kemal was heir, and which he did his best to realize by doing away with caliphate and sultanate, by changing the alphabet and the headgear, by seeking methodically to uproot the institutions of Islam and banish its influence. Thus we see an article (174) enumerating those enactments extending over the years, from 1924 to 1934, which 'aim to raise Turkish society above the level of contemporary civilization, and to safeguard the secular character of the Republic', and affirming that nothing in the Constitution may be construed or interpreted as rendering them unconstitutional. These laws relate to 'the Unification of the Educational System', the 'Wearing of Hats', the 'Closure of Dervish Convents and Tombs', the 'principle of Civil Marriage', the 'Adoption of International Numerals', the 'Adoption and Application of the Turkish Alphabet', the 'Abolition of Titles and Appellations such as Effendi, Bey, or Pasha', and the 'Prohibition of the Wearing of Certain Garments'. This article is a repetition of the identical article in the 1961 Constitution. The history of the Turkish Republic's engagement with the religious issue during six decades, and the failure of secularism to eradicate the traditional attachment of the mass to Islam makes this repetition both comprehensible and significant. As significant and instructive are two other articles concerned with religion. Article 136 insists that the Department of Religious Affairs shall exercise its duties 'in accordance with the principles of secularism', and Article 24 forbids the exploitation of religion for purpose of political influence and, it adds, 'for even partially basing the fundamental, social, economic, political, and legal order of the State on religious tenets'. This statement occurs in an article concerned, among other things, with religious instruction in schools. This takes place, the article stipulates, 'under State supervision and control'. The sentence which follows is therefore the more surprising: 'Instruction in religious culture and moral education' it declares, 'shall be compulsory in the curricula of primary and secondary schools'. 'Religious culture' is obviously a laborious circumlocution for Islam. To Atatürk such a constitutional provision would have been inconceivable, while the Democrats who did so much in the 1950s to make it possible for the mass to follow its Islamic bent, would never have dared to propose it as an addition to the then constitution. That it should now find a place in a constitution which proclaims Turkey to be a secular republic indicates that the tendencies and preferences given a relatively free rein in the three decades following Menderes' assumption of power can no longer be suppressed or ignored by the official classes who stand guard over Atatürk's legacy.

One feature of Turkish politics after 1961 which was believed to promote instability was proportional representation which was introduced then, and which allowed the growth and proliferation of small parties which held the balance of power in the Assembly, demanded a high price for their co-operation, and prevented the pursuit of coherent long-term policies. The electoral system was considerably modified following the ultimatum of March 1971. The new electoral law, promulgated in June 1983, modified it still further so as to make the election of members from small parties even more difficult. Political parties which obtained less than 10 per cent of the vote were excluded from seat allocation; also excluded were parties which obtained less than the average number of votes necessary for the allocation of a seat in an electoral district. In 1987 the electoral arrangements were made even less favourable for small parties: in those constituencies which returned six deputies (which was the maximum) the average number of votes that a party has to obtain in order to be allocated a seat was obtained by dividing the number of votes cast by five (and not by six); also, major metropolitan districts such as Istanbul, Ankara, Izmir, and Adana were divided into smaller and less populous electoral districts, further reducing the chances of minor parties and thus of the proliferation of small parties.

The political parties law promulgated in April 1983 banned from party activity for ten years the leading members of the political parties that had existed at the time of the *coup d'état*, and that had been already closed down in 1981; all those who had been members of the Grand National Assembly before 12 September 1980 were banned from active politics for five years; and the law forbade the formation of new parties that adopted the names and emblems of those that had been banned. Parties proposing to contest elections, and the candidates they were putting up, had to obtain the approval of the National Security Council. Fourteen parties hoped to take part in the elections, but only three were approved: the Populist Party, led by a former official; the Nationalist Democracy Party, led by a retired officer; and the Motherland Party, led by a former deputy prime minister who had served in the administration set up by the National Security Council after the *coup d'état*. Of the candidates proposed by the three parties for the November 1983 elections, 392 from the Motherland Party were vetoed, 398 from the Nationalist Democrats and 389 from the Populists.

The leader of the newly formed Motherland Party was Turgut Özal whom Demirel had appointed in 1979 head of the State Planning Organization, and who was the author of a radical policy for financial stabilization which was necessary if Turkey was to receive further help from its foreign creditors. After the *coup d'état*, he was appointed as a deputy Prime Minister, again in charge of the economy. In July 1982 Özal resigned his post and the following year formed the Motherland Party. The new Party was an amalgam of former supporters of the Justice Party, the National Salvation Party, and the Nationalist Action

Party. In the electoral campaign the generals' preference went
to the Nationalist Democracy Party, and President Evren
himself intervened in its favour in the last stages of the
campaign. But neither this Party nor the Populists came near
to winning. The Motherland Party gained over 45 per cent of
the votes and 211 seats, the Populists 30.5 per cent and 117,
and the Nationalist Democracy Party, in spite of its endorse-
ment by Evren, 23.3 per cent and 71 seats. That the Motherland
Party may have been perceived as the natural continuation of
the Democratic and the Justice Parties in going with the grain
of Turkish society probably explains its success, as does the fact
that Özal was known to have been a member of the National
Salvation Party, for which he stood as a candidate in the 1977
elections, and in which a brother of his was an influential
member. He had also acquired a reputation, while a high
official and a minister, for promoting policies to combat inflation, the ravages of which sooner or later were felt by the great
majority.

When Özal took office, he was indeed firm in his public
professions against following inflationary policies. 'First' he
said, when submitting his Government's programme to the
Grand National Assembly in December 1983, 'inflation must
be curbed. The supply of money will be carefully controlled
according to the course of the economy.' Budget deficits would
be kept to 'a rational level', and intervention by the state in
industry 'the cost of which always falls back on the low and
fixed income groups, will be avoided'. It seems, however, that
in a country like Turkey it is very difficult for a political leader
to forgo the temptation of printing money, which provides a
quick and easy means of dealing with economic difficulties in
the short run. The fact remains that, under Özal, inflation,
having fallen to 24 per cent in 1982, quickly climbed up again
and in 1984 was 53 per cent and has since remained very high.

In 1984 a Social Democratic Party, heir to the Republicans,
was formed under the leadership of Erdal İnönü, son of
Atatürk's successor as President, and until 1972 leader of the
Republicans. The following year the Populists, who had had
such little electoral success, combined with them to form the

Social Democratic Populist Party. Also in 1984 those of Demirel's followers who had not joined the Motherland Party formed the True Path Party, which Demirel began leading again when the ban on his taking part in politics was lifted in 1987.

The second general election under the new constitution took place in November 1987. The Motherland Party took 36.3 per cent of the vote and 292 seats, the Social Democratic Populist Party 24.7 per cent and 99 seats, and the True Path Party 19.1 per cent and 59 seats. Özal therefore continued as Prime Minister. In November 1989, Evren's term as President expired, and as the Constitution required, it fell to the National Assembly to elect a successor. Given the Motherland Party's majority in the Assembly, it was a foregone conclusion that Özal should be elected. His successor as Prime Minister was Yildirim Akbulut, formerly Speaker of the Assembly.

Özal is thus the first President of Turkey who is a civilian, and not a military officer recently, or not so recently, retired. He and his Party, however, have come to power under the auspices of the armed forces who are the true authors of the Constitution and of the laws on elections and on parties, thanks to which Özal and his Party have continued to govern with the support of stable majorities—a far cry from the conditions ruling in the 1960s and 1970s. It remains true, however, that the regime created by army leaders, continues to be under their tutelary protection. The armed forces consider themselves, indeed, the protectors and guarantors of the constitutional order. It must remain on the cards that, should conditions once more call for intervention, they will feel justified in intervening.

In the light of seven decades' attempts to set up a modern constitutional state—attempts which have by and large failed— it would perhaps be unwise to count on the longevity of this latest constitutional arrangement. There are at least three reasons for caution. The first is the standing, powerful temptation on the part of the rulers to print money. Time and again successive rulers since Menderes have resorted to it, and they found the economy inevitably breaking in their hands, with all the consequent popular discontent and disaffection. The same electorate, for example, which gave Özal his majority in

November 1987, placed the Motherland Party third in the local elections of March 1989, with only 22 per cent of the vote, owing, it is believed, to the disappointment at his failure to fulfil expectations.

In the second place, it is a moot point whether the political parties are really willing to recognize one another's legitimacy and *bona fides*—a necessary condition if representative, parliamentary government is to be carried on. Thus, the two opposition parties represented in the Assembly boycotted the presidential elections of November 1989, claiming that the majority on which the Motherland Party relies is deceptive, since it enjoys the support of only 36 per cent of the votes—a proportion which, they claim, has recently diminished greatly. The leaders of the True Path Party and the Social Democratic Populist Party refuse to accept that President Özal embodies the unity of the nation, and will have no dealings with him.

There is, finally, the religious issue which has remained unresolved ever since the beginning of the Republic. The successive constitutions have reiterated that secularism is one of their fundamental governing principles, and they are all of them completely unambiguous in prohibiting any connection between religion and politics, or any undoing of the Kemalist measures which had attempted to transform Turkey from an Islamic into a modern European state. And yet, the mass of Turks have remained profoundly attached to Islam, as was shown whenever a government has allowed them to give free rein to their sentiments. Witness to this is the success at the polls of the Democratic Party, its successor the Justice Party, and later on, of the Motherland Party—a success which can be attributed, even if only in part, to the voters sensing that these parties are sympathetic to Islam and as helpful in promoting its institutions and its interests. On the other hand, the intellectual and official classes, including the leadership of the armed forces, are predominantly wedded to Kemalism, and liable to react with hostility when they suspect any tampering with Kemalist tenets. For instance, *12 September in Turkey: Before and After*, mentioned above, which was officially published by the National Security Council, is as severe in its judgements concerning the Islamists, the National Salvation Party, and its leader Necmettin Erbakan as on the left-wing groups and Turkeş and his Nationalist Action Party. And yet the Prime Minister, now President, is a former Salvationist whose sympathies with the Islamic current are evident and tangible. When he formed his administration in November 1982, he appointed as Minister of Education Vehbi Dinçerler, who was said to be a member of the Naqshbendi sufi order, who decided to ban the teaching of Darwin's theory in primary and secondary schools, to include Arabic in the secondary-school curriculum, to include Islamic publictions as suggested reading for primary and secondary school students, to have school textbooks rewritten in an Islamist sense, and to dismiss the director of a lycée in Izmir and two of his teachers for opposing the building of a mosque on school grounds. Özal's government indeed decided that mosques were to be built within the Grand National Assembly, in various ministries, and on university campuses. Özal himself has backed the religious group within the Party who had formerly been Salvationists and who were led by Mehmet Keçeciler. Keçeciler has become a minister of state in Akbulut's administration, his elevation having apparently been previously blocked by Evren while President. A newspaper report shortly following Özal's election to the Presidency declared that he is known to be in favour of the headscarf for women, and that when attending Friday prayers, as he regularly does, 'lines of citizens wait to kiss his hand'. Also, the Welfare Party—a reincarnation of the National Salvation Party, and also led by Erbakan—has hailed Özal as the first President of Turkey 'whose forehead touched the prayer mat'.[13] The tension between secularism and Islam present in Turkish politics from the foundation of the Republic, is, thus far from subsiding, let alone disappearing, and in certain circumstances might prove too strong for a still-fragile constitutionalism.

However, another reading of the prospects of Turkish politics is possible. In an essay published in 1988, Professor Dankwart Rustow argues that from 1950 on schools have multiplied, literacy has increased, newspapers have multiplied, urbanization has been uninterrupted. This has transformed, and will go on transforming, Turkish politics, as political parties competing for the votes of illiterate peasants will have to offer more schools, better roads, and other amenities. This in turn will facilitate mass migration from the countryside in search of greater economic opportunities. Both politicians and voters will increasingly act according to the logic of democratic competition. Turkey, he also argued, does not suffer from ethnic problems which are naturally insoluble. The political differences have been between rich and poor; they are over economic policy, over whether agriculture or business is to be more favoured, over civilian–military relations, and, he concludes, 'those divisions have been attenuated by the processes of economic development and social mobility which democracy itself has done so much to accelerate'.[14] Only time, of course, will show if this optimism is warranted.

5

Constitutionalism and its Failure: II

1. Parliamentary Government in Egypt 1923–52

THE political history of Turkey after the establishment of the Republic was, as may be seen from the preceding Chapter, extremely checkered. Time and again a constitutionalist mode of government broke down and the break-down was signalled by a military intervention or *coup d'état*; but also, time and again, the authors of these military initiatives rejected the temptation of themselves assuming the responsibilities of government, and persisted in leaving, or speedily transferring back, these responsibilities to elected representatives and their political leaders. In so cleaving to constitutionalism, the political and official classes of Turkey showed themselves faithful to a tradition inaugurated by the Young Ottomans in the 1860s, and continued by the Young Turks, whose first action following the success of their *coup d'état* in July 1908 was to demand the re-establishment of the 1876 Constitution. As we know, constitutionalism failed totally in the Ottoman Empire and it is all the more remarkable, in the light of this failure, that the founders of the Republic—military men imbued with the hierarchical values which a military career necessarily inculcates—should have remained steadfastly faithful to this ideal, and should have successfully transmitted faith in it to their successors, even though there was little evidence to sustain this faith.

The constitutionalist ideal was, of course, derived from Europe and inspired by the belief that in it lay the secret of European superiority. European states, however, neither pressed, nor indeed were in a position to press, for the establishment of parliamentary government. The driving power, from first to last, came entirely from within. The same was, of course, true of the constitutional revolution of 1906 in Persia—a revolution which again proved a failure and that within the space of a few years. But the situation was very different in Egypt. Early in the War Egypt had been declared a British protectorate, when the tenuous and purely formal Ottoman suzerainty was ended. In token of this, the ruler of Egypt, a descendant of Muhammad Ali Pasha, exchanged the title of Khedive (which had been bestowed by the Ottoman Sultan on Ismail Pasha, who ruled Egypt between 1863 and 1879) for that of Sultan. But whether Khedive or Sultan, the ruler of Egypt, whose power over the country and its inhabitants knew in theory no limits, became in reality subordinate, after the occupation of 1882, to the British representative, who controlled a country-wide network of British advisers and inspectors supervising every aspect of Egyptian administration. After 1882, in brief, British views and policies were paramount in Egypt.

Shortly after the occupation, in 1883, the Khedive Tawfiq, who had succeeded his father in 1879, established a Legislative Council and a Legislative Assembly. Both were small bodies, of which a substantial part was appointed. The members met in private and had few, if any, powers. In 1913 the Council and the Assembly were merged to form a new Legislative Assembly in which seventeen members were nominated by the Government and sixty-six elected by indirect suffrage. The new Assembly met in public, but, belying its name, it had a predominantly consultative function. The new-style Assembly was elected and met during 1914, but the War caused it to be suspended in 1915, and it did not meet thereafter.

The War and its aftermath indeed caused a great change in the situation of Egypt, a change which was fundamentally affected by British policy, and which launched the country into some three decades of parliamentary and party politics. In 1917 Fuad, a son of Ismail, quite unexpectedly became Sultan. His nephew, the Khedive Abbas Hilmi, happened to be in Istanbul when the Ottomans joined the Central Powers in November 1914. The British authorities, mistrusting his loyalty and fearing his ambition and intrigues of which they had had reason to complain for many years, deposed him and installed as Sultan an uncle of his, Husayn Kamil, who proved friendly and pliant. When he died, it seemed to the British authorities that, on balance, Faud would be the most suitable successor. They proved badly mistaken, since shortly after he became Sultan and thereafter, until his death in 1936, Fuad showed himself eager for self-aggrandizement, and both wily and tenacious in pursuing his aims.

The stresses of war created discontent in the country. Military expenditures brought with them inflation which bore most heavily on the poor. Imports became scarce and this helped to increase prices further. To palliate the scarcity of foodstuffs, farmers were compelled to change from the lucrative cultivation of cotton to the growing of food. The armed forces demanded the requisitioning of horses and mules, and what amounted to a conscription of fellahin in order to man a labour corps used for the digging of trenches and the like in the theatre of war. Both demands created great discontent and fear, the more so that the requisitions and the choice of those designated to serve in the labour corps fell into the hands of local notables and *omdas*, who were believed, generally with reason, to be arbitrary, greedy, and corrupt. The great justification of the British control of Egypt had always been that it ensured just rule, protecting the small man from the insolence and rapacity of wealth and office which the record of native Egyptian government, from the Mamelukes to Muhammad Ali and his successors, amply showed that otherwise nothing could curb. The stresses of war, however, caused a slackening in the British apparatus of administrative control and allowed abuses to become rife. To the demands of the British army, represented as both necessary and urgent, which provided opportunities for the abuse of power and influence, a British High Commissioner, his advisers, and agents could not in the circumstances offer determined resistance—at any rate, they did not do so.

Egypt then was, by the end of the War, in a dangerously volatile state. The extent of the discontent neither the High Commissioner nor his masters in London seem to have appreciated. Also unexpected by the authorities in London was a step taken by some Egyptian political leaders, it would seem in conjunction with Fuad, or at his urging—a step which was to lead to an explosion of these pent-up discontents, the consequences of which were, within a few years, to change radically the governance of Egypt and its political landscape. Two days after the armistice of 11 November 1918, the High Commissioner, Sir Reginald Wingate, received, at their request, three prominent figures, who, however, then held no public office, and who asked for permission to go to London in order to discuss with the authorities there the conditions of the Protectorate which had been proclaimed in 1914. They were acting in concert with Fuad and his ministers. Fuad, as Wingate had earlier reported, was dissatisfied with the Protectorate. He was also uneasy about his position, threatened as it might still be by his nephew, the deposed Khedive, whose claim to rule was still alive. When President Woodrow Wilson published his Fourteen Points, Fuad was prompt to welcome them. By raising the issue of the Protectorate, Fuad calculated that he would increase his standing in the country, and also perhaps eventually enjoy powers similar to his father's before the days of European interference in Egyptian affairs. As for Fuad's ministers, they had been in office all through the war, and could thus be accused by their opponents of having put British interests before Egyptian. If Fuad and this unofficial delegation were raising the issue of the Protectorate and implicitly asking for Egyptian independence, then those ministers could not afford to lag behind.

The most prominent member of the delegation was Sa'd Zaghlul. By 1918 he had had a long and varied public career. As a young man he had attended the Azhar university-mosque. He became literary editor of the official *Egyptian Gazette* when Tawfiq became Khedive in 1879, and was implicated in a minor way with the officers' movement which followed. He then took up law and prospered in it. The editor of the *Gazette* had been Sheikh Muhammad Abduh who was far more deeply implicated in 'Urabi's movement and was in consequence sentenced to exile. Lord Cromer, who in effect ruled Egypt from the British occupation until 1906, had great regard for Abduh as a moderate and progressive Muslim, and had him recalled from exile. He was also given high office in the administration of Egyptian *shari'a* courts and in religious education, and eventually appointed Mufti of Egypt, one of the two highest religious posts in the country. Zaghlul became known as one of his followers—those followers whom Cromer described in his Annual Report on Egypt for 1905 as 'the Girondists of the Egyptian national movement', and from whom he hoped so much, in reforming Islam and making it compatible with modern civilization. Zaghlul married the daughter of Mustafa Fahmi Pasha, who served for many years as Minister and Prime Minister and who was considered pro-British. This connection, together with Zaghlul's relationship to Abduh, led Cromer to further his career. He was appointed Minister of Education in 1906, and served in this post until 1910, when he became Minister of Justice. He was identified throughout with Islamic liberalism and with sympathy for constitutionalism. His record thus indicates that, so far, his political stance was pro-British and against the Khedive who, as is well known, was unfriendly to Cromer and his local supporters. In 1912 Zaghlul in fact clashed with the Khedive and resigned his office. He then coquetted with the group known as the Nationalist Party, which had been promoted by the Khedive, but with whom they had fallen out. Shortly afterwards, however, he became a partisan of the Khedive. He was elected to the new-style Legislative Assembly where he was chosen as its Vice-President. In the Assembly he became the defender of the Khedive's causes, and made a name for himself as a powerful debater. A high British official wrote at the time that he was 'the dominating personality throughout the [1914] session, and he has the makings of a successful demagogue'. As has been said, the outbreak of war led to the deposition of Abbas Hilmi and the prorogation of the Assembly.

Known now as an opponent of the British connection, Zaghlul was in the political wilderness. He remained for a time true to the cause of the deposed Khedive. He told the British Judicial Adviser in 1917 that if the Assembly were convoked one would have to swear allegiance to Abbas Hilmi's successor, and he himself could not do so, since upon his election he had sworn allegiance to the Khedive.[1]

Zaghlul, then, had up to this point executed quite a few political somersaults. He now executed yet another which proved the most important in his career—which was in fact to transform it out of recognition. This unemployed politician now approaching his sixties was shortly to become, in a manner no one could have foreseen, a powerful leader inaugurating a new kind of politics in Egypt. It must have been shortly after his conversation with the Judicial Adviser that Zaghlul changed his allegiance once more and became a confidential adviser to the new Sultan who, it will be remembered, had just assumed office and was fearful of the ambition of his deposed nephew now hovering in the wings. At the end of 1917, Fuad asked that Zaghlul should be made a minister. The British decided, on balance, that even though the oratorical powers Zaghlul had shown during the brief life of the Legislative Assembly might be tamed by office, it would still be preferable not to have him inside the Egyptian administration. Fuad's request was turned down, and so it was that when Zaghlul went to see Wingate a year later as the head of a delegation, it was as a prominent personality, but one free of official responsibilities.

The visit to Wingate, as has been said, was concerted with Fuad and his ministers. Wingate referred the request to London which immediately forbade any Egyptian personalities, whether official or unofficial, to leave Egypt in order to go either to London, or to Paris where the Peace Conference was meeting. The Egyptian ministers immediately resigned. There was little else they could do since, given their acquiescence in the demands of the British army during the war, had they done otherwise they would have been open to damaging attacks, particularly as they were by no means sure that Fuad would not himself work to undermine them. Wingate persuaded the ministers to stay their hand for the time being. Zaghlul was now in the limelight, he and his fellow-delegates being acclaimed as the standard-bearers of the Egyptian cause. Wingate went to London in January 1919 and tried to persuade the Government to be less negative in their response to the requests which had been made. The Government did in the end say that some time in the future they might discuss the issue of the Protectorate with the Egyptian ministers, but ruled out Zaghlul taking part in such discussions. This was unacceptable to the ministers, since had they engaged in discussions with the British Government without Zaghlul taking part, they would have been open to the accusation that they had not stood up for Egyptian interests with sufficient vigour. Their resignation then became final.

Following his rebuff in November 1918 Zaghlul had not been idle. Petitions asking that he and his fellow delegates should be allowed to travel to Europe in order to publicize the Egyptian case were spread throughout the land, and attracted large numbers of signatures. This could not have been done without a great deal of organization, and without the help, or at least the acquiescence, of the royal palace as well as of the ministers. On the instructions of the British Adviser to the Ministry of the Interior, provincial authorities attempted to confiscate the petitions. Zaghlul protested to the Prime Minister, Rushdi Pasha, who asked the Adviser what answer he should give. Very maladroitly, he was told to say that this was being done on the Adviser's orders. The Adviser's answer was published, and it thus became apparent that in this dispute the British were ranged on one side, while all the Egyptians—Zaghlul, the administration, the Sultan—were on the other. Of all these parties Zaghlul proved the most able to exploit the situation to his advantage. His name was now to the fore as the leading champion of Egyptian rights. When the ministers resigned following the British refusal to allow Zaghlul to take part in any future discussions, he moved quickly to prove that he was now the arbiter of Egyptian politics. On 5 March 1919, he went to the palace at the head of a delegation and left a letter drawn in the most insolent and threatening terms in which he warned Fuad against appointing anyone to replace the resigning ministers. The letter spoke of Fuad accepting the sultanate 'during the temporary and illegal protectorate', declared that 'the nation' was asking Fuad for help in attaining independence, 'however much this might cost Your Highness', affirmed that no 'honourable and patriotic' Egyptian would accept to take the place of the ministers and thus work against 'the will of the people', and it advised Fuad that before taking any action he should ascertain the opinion of the 'nation'. Zaghlul thus turned the tables on his late patron, catching him with his own stratagems. Fuad took fright. He appealed for protection to the Acting High Commissioner who was in charge during Wingate's absence. Most chivalrously and most foolishly this official—Sir Milne Cheetham—hastened to the defence of the Sultan—who had himself set in motion the machinery now threatening him. On 6 March, Cheetham advised that Zaghlul and his friends should be interned outside Egypt 'for the sake of the Sultan's prestige, which is a political interest to us'. On 7 March the Foreign Office, without consulting Wingate who was still in London, signified its agreement, and on 9 March Zaghlul and three companions were arrested and sent to Malta.

The volatile state of Egypt and the agitation which had gone on for some four months immediately led to the eruption of widespread disorders. It did not take more than a few days of rioting to panic the Acting High Commissioner who now, on 15 March, urgently suggested that Zaghlul and his friends should be allowed to go to France and England—the very thing which had been for months consistently denied them. It was quite clear to his superiors, to the Prime Minister, Lloyd George, to Balfour, the Foreign Secretary (who were then both in Paris), and to Curzon, in charge of the Foreign Office in London, that Cheetham was unable to cope. Wingate was still in London, but instead of asking him to go back and take charge, Lloyd George and Balfour decided to appoint Lord Allenby, the conqueror of the Levant, as 'special high commissioner'. The reasons for Wingate's supersession are obscure. Wingate had a long experience of the Middle East, and understood Egyptian politics and personalities. But he had aroused jeaousies and enmities in the Foreign Office, and he was, unjustifiably, accused of weakness and misjudgement in agreeing to receive Zaghlul the previous November. Allenby was a soldier, and Lloyd George hoped that, as such, he would know how quickly to put an end to the disturbances. Allenby, however, as speedily became apparent, was out of his depth in Egyptian politics, and had no feel for them. Instead of leading Egypt back into calm, and preserving the British position—the purpose for which he was appointed over Wingate's head, he insisted on taking action which was hasty, ill-judged, and which led to inextricable complications, and launched Egypt into a long period of political instability.

Allenby reached Egypt on 25 March 1919. The disturbance unleashed by Zaghlul's banishment had by then been largely contained through the resolute action by the commander of the British troops. On 31 March Allenby proposed that Zaghlul and his friends should be released from Malta and allowed to go to Europe. The proposal appalled Curzon, but the Prime Minister did not dare to dismiss Allenby who had been so recently appointed. On 14 April, Allenby sent yet another telegram demanding a favourable reply to his request, and two days later announced that he was setting free Zaghlul and his companions, who proceeded to Paris. Allenby thus committed a great blunder. He had thought that his action would be a master-stroke, showing the Egyptians that he combined clemency and firmness. In fact the Egyptians took it to be a sign of weakness that someone who defied British authority should unconditionally go scot-free.

Allenby hoped that with Zaghlul free, it would be possible to get the ministers who had resigned a few weeks before once more to take office. Egypt was not a Crown colony which could be directly administered by British civil servants. It was a protectorate where the native government was indispensable for the running of the country, even though in everything that mattered the last word lay with the High Commissioner and the British advisers. Allenby thought of these ministers as moderate. This was no doubt true in general and in the abstract. However, in the situation which he himself had engineered 'moderation' became irrelevant. Zaghlul was now in Paris, claiming to be the standard-bearer of the struggle against the British Protectorate. This struggle, he would claim, was a popular struggle, as witness the great support which his approach to Wingate had elicited, and the uprising which had followed his banishment. He was unfettered by the responsibilities of office, and in Paris he was beyond the reach of the British authorities in Egypt. He could therefore make whatever demands he wanted, and appeal to the people over the heads of both ministers and Sultan. Zaghlul now raised the banner of 'complete independence'. He was, he now claimed, the only legitimate spokesman of the whole Egyptian people and the head of a nationwide movement devoted to this object. The movement he called the Wafd, i.e. Delegation—a title which reminded this audience of the delegation which had visited Wingate and so had struck the first blow for Egyptian independence. That this was the case Zaghlul proved from the fact that on his banishment the whole Egyptian people had risen in revolt. In the face of such a claim neither the Sultan nor any lesser figure could afford to pitch their own demands any lower, or publicly to criticize Zaghlul and his overweening stance, even though his activities threatened their own position. This situation graphically exhibited the consequences of Allenby's miscalculation.

Shortly afterwards, Zaghlul's position was to be considerably fortified. In order to deflect Allenby from his decision to release Zaghlul in March 1919, the Government had suggested that instead, a committee be sent to enquire on the spot into Egyptian grievances. Allenby agreed that this might be done, but did not accept this as a substitute for what he was proposing. The Government's choice fell on Lord Milner, the Colonial Secretary, as the head of this committee. Milner seemed an ideal appointment since he knew Egypt well, having served there as an under-secretary in the Ministry of Finance and published a book, *England in Egypt*,[2] which described the reforms and improvements necessary to put Egypt on a sound footing, and argued that the British presence in Egypt was justified by the necessity of carrying them out. Milner was at the outset reluctant to go to Egypt. Towards the end of April, in a letter to Curzon, he spoke of 'the great blunder' made by Allenby in releasing Zaghlul simply in order to be able to form a government. It was, he said, 'a most ill-fated policy of concession'.

For thirty years we have governed Egypt because of the conviction in the minds of the intriguing Pasha class that in the last resort we could and would do without them. That conviction, for some reason or other, has been weakened of late years. Hence the present troubles, which are simply a 'try on' on the part of the caste.

For this reason it would be a mistake to send a mission of enquiry just then:

It looks as if we were flustered, afraid of the situation created by the non-existence of an Egyptian Ministry and the naked assertion of British authority, and felt that something must be done to get us out of a hole.[3]

Milner thus decided that it would be better if his Mission—as the committee came to be known—did not visit Egypt in the immediate aftermath of the disturbance. In the event, it did not go until December 1919. Zaghlul, of course, considered the Mission and its activities as inimical to his own position. He held that as he was the only legitimate spokesman of the Egyptian people, the Mission should direct its enquiries exclusively to himself. He therefore decreed that no one in Egypt should talk to Milner and his colleagues, that, in short, the Mission should face a total boycott, which would persuade it that the solution to the problem it was investigating lay solely in negotiation with himself. This in itself made Egyptian political figures unwilling to have any public commerce with the Mission. Also, the delay in the arrival of the Mission allowed public agitation by Zaghlul's supporters to be orchestrated, and for intimidation by a terrorist apparatus forcibly to make Zaghlul's point for him. This apparatus seems to have grown from a nucleus organized before the war by Egyptian nationalists who believed that the British occupation could be ended in this way. The murder in 1910 of the Prime Minister, Butrus Ghali (a Copt), whom these nationalists accused of collaboration with the British, was almost the only action which the terrorists undertook, but it did not succeed in shaking the British hold on Egypt, or even in modifying British policies. Now Zaghlul somehow or other became connected with this apparatus and came to direct its activities. In the disturbed and effervescent state of Egypt after the war, and especially follow-ing Zaghlul's release from Malta, widely taken to be a sign of British weakness, the activities of this apparatus obtained a much greater resonance. The intimidation for which it was responsible succeeded in spreading widely the impression that no one in Egypt would have anything to do with the Milner Mission because the Wafd had forbidden any contact with it.

Before he arrived in Egypt, Milner considered his task was to negotiate a treaty between Egypt and Britain which would secure to the Protecting Power, fortified as it had been by the official recognition of the Protectorate by the Paris Peace Conference, in spite of Zaghlul's representations, those powers of control over the Egyptian Government which were abso-lutely necessary. However, during his stay in Egypt Milner came to be persuaded by the boycott and the clamour and abuse which dogged its footsteps that Britain should no longer take responsibility for the good government of Egypt, but should simply strive to secure its own strategic requirements, safeguard imperial communications, and protect the interests of British residents in Egypt. He was fortified in this change of view by the influence of two members of his Mission, J. A. Spender, a Liberal journalist, and Sir Cecil Hurst, the Legal Adviser to the Foreign Office. What they thought may be gathered from remarks of theirs which occur in the minutes of the Mission. They had been hearing evidence about pressure and threats used by Zaghlulists in order to force village *omdas* to collaborate with them:

Mr Spender and Mr Hurst observed that the principle might be adopted of allowing the Egyptian government to do things shocking to us as long as they did not affect foreign interests.[4]

The change in Milner's attitude which occurred during his stay may be gathered from a conversation between him and a European acquaintance. This acquaintance, a businessman long resident in Egypt, maintained that the welfare of Egypt required more, not less, British control. Milner admitted that such control was in the interest of the masses. But these masses, he said, were mute and the only clamour to be heard was that of politicians abusing and reviling the British. In the circumstances, what his acquaintance was proposing was 'a fine ideal, but I cannot say that I feel convinced that Great Britain would have the power or the will to pursue it'.

On leaving Egypt at the beginning of March 1920, the Mission unanimously adopted some General Conclusions which embodied the views of Milner, Spender, and Hurst. They declared that the British Government 'should be guided by the principle to restrict the direct exercise of British authority to the narrowest possible limits'. This object was to be assured by an Anglo-Egyptian treaty which would define these limits and simultaneously 'define the general character of the future constitution of Egypt'. Such a treaty, duly negotiated, would then be ratified by a representative assembly. The General Conclusions, however, recognized that 'owing to the backwardness of the mass of the people, of whom 90 per cent are quite illiterate, it will be many years before any elected Assembly is really representative of more than a comparatively limited class. Parliamentary government under the present social conditions', the General Conclusions recognized, 'means oligarchical government, and, if wholly uncontrolled, it would be likely to show little regard for the interests of the Egyptian people'.[5]

A treaty would have to be negotiated. With whom? In the year and a half since he had visited Wingate, Zaghlul had been allowed by Cheetham's and Allenby's mistakes, and by unwillingness of any Egyptian personality to challenge, let alone outface the exile, to set himself up as the sole legitimate representative of Egypt, without whom no negotiation could progress and no agreement hold. The Egyptian figures with whom the British could conceivably negotiate were too afraid for their lives and positions to do anything of the kind. All that they were willing to do was to act as intermediaries between the Protectorate and its challenger, and hope to derive from this a broker's commission, or perhaps even larger profits: if Zaghlul or the British faltered or made a false move, the broker would become principal and might, with luck, claim the credit for ending the Protectorate. One of these figures, Adli Pasha, who had served in the war-time ministry which resigned on

Zaghlul's banishment, persuaded Milner that he should negotiate with Zaghlul, and offered to go to Paris in order to facilitate
matters. Milner, here going beyond his remit, without inform-
ing the Cabinet and obtaining their consent, agreed to do so.

In these negotiations Milner showed extraordinary ineptitude
and allowed himself to be consistently outmanœuvred by Zaghlul. Persuaded by Adli and a British adviser, Osmond Walrond,
on whom he relied a great deal but who had little judgement,
Milner invited Zaghlul to talks in London. The invitation was
for private talks, to Zaghlul in his personal capacity, but
Zaghlul hastened to publish that the Milner Mission had invited
the 'Egyptian Delegation' to London in order 'to discuss the
bases of an agreement' between Egypt and Great Britain, and
that colleagues were going to London ahead of him to 'make
sure of British intentions regarding Egyptian aspirations to
complete independence'. This is not at all what Milner had
intended, but he was afraid to disown Zaghlul's claim for fear
that this might endanger negotiations. Zaghlul's prestige in
Egypt went even higher. In the course of the negotiations,
Milner also forbade the Cairo Residency to make plans to take
action against the Zaghlulists in case negotiations broke down,
and this in spite of much evidence that the Wafd was using
high-handed methods for extracting contributions in the
countryside. He also ordered a delay in the arrest of the man
suspected of heading the Wafd's terrorist organization and in
searching his premises, because he feared that it might uncover
documents embarrassing to Zaghlul. By the time permission
was extracted from Milner it was too late, and the search
produced nothing. There was, however, sufficient evidence to
convict the suspect, Abd al-Rahman Fahmi. He was found
guilty and sentenced to death. Zaghlul protested, and Milner
tried hard to have the conviction quashed, but the Judicial
Adviser to the Egyptian Government, to whom the records of
the trial were submitted, concluded that the trial had been
regularly conducted and the evidence adequate, and could find
no grounds for quashing the conviction. The death sentence,
however, was commuted to fifteen years' imprisonment. The
prisoner, also, was allowed uncontrolled visits and uncensored

post and took up quarters in the prison hospital. Milner also refused to allow further investigation of other suspects because Allenby had reported that after the arrest and conviction of Abd al-Rahman Fahmi, terrorist outrages had ceased. If this was so, Milner argued, better let sleeping dogs lie.

Milner's negotiations with Zaghlul in London lasted from the end of May until July 1920. In them Milner conceded recognition of the independence of Egypt as a constitutional monarchy with representative institutions, Egypt in exchange ackowledging Great Britain's right to protect the privileges of foreigners resident in the country, and safeguard its strategic interests and imperial communications. This was the so-called Milner–Zaghlul agreement, except that it was no agreement since, in the end, Zaghlul said that he could not commit himself to the terms which had been negotiated without first obtaining the approval of the Egyptian people, but he assured Milner that he would send emissaries to commend the agreement to the people. Zaghlul did send emissaries, avowedly with this object, but he also sent a secret letter to his followers, explaining that whatever the emissaries might say in public, he himself was not in favour of the agreement and would accept nothing less than complete independence. Zaghlul also let it be known that but for Adli, who had proved a 'disaster', he himself would have obtained much better terms. Not only did the broker, then, not gain a commission, but he was left to carry the can for the failure of the transaction. Zaghlul proved much wilier than Adli.

It was only when his policy was in ruins that Milner's actions were disclosed to the Cabinet, in October 1920. What is more, Curzon, the Foreign Secretary, had, through the months of negotiation with Zaghlul, chosen not to inform himself of what was going on. But though the agreement failed, it had significantly damaged the British position. As Winston Churchill was presciently to observe shortly afterwards, Milner's concessions would prove to be the minimum from which any future Egptian negotiator would feel obliged to start. Milner's failure returned the initiative to Allenby in Cairo. He now demanded to know what the Government proposed to do in order to settle the issue. On the one hand, he reported that the Sultan was asking for a public declaration of British policy, and on the other he pointed out that any retreat from Milner's offer would lead to a resurgence of Zaghlulist agitation. Either the Government should immediately accept Milner's proposals or it had to declare that the protectorate was 'not a satisfactory relation in which Egypt should continue to stand to Great Britain'. At Curzon's urging, this latter line was adopted. A declaration to this effect was issued on 22 February 1921.

This was a signal weakening of the British hold on Egypt for which Zaghlul would—and did—claim the credit. But in justice, credit was also due to Allenby's bungling and Milner's remarkable innocence. The declaration of 22 February conceded, with nothing, however, obtained by Britain in return, that Zaghlul, Fuad, and his ministers were justified in questioning the Protectorate and implicitly at first, and later on more and more explicitly, claiming the right to complete independence. If the Protectorate was not a satisfactory relation between Egypt and Great Britain, then it had to be replaced—replaced, as the Milner Mission had proposed, by a treaty. Negotiation of a treaty with Zaghlul had been a mistake on more than one count—one, particularly crucial, was that though he claimed to be the spokesman of the Egyptian people, yet he had in reality no authority to speak on behalf of the Sultan and the Egyptian Government, the only entities empowered to enter into international agreements and treaties. This anomalous proceeding was justified in the Milner Mission report (published in the spring of 1921) by the argument that a treaty would have to be ratified by a 'genuinely representative' assembly, and the Zaghlulists were bound to command 'a substantial if not an overwhelming majority' in such an assembly. The argument shows that on this as on other issues, the Milner Mission had utterly fallen for Zaghlul's claims. A substitute had now to be found for Zaghlul. Adli, as has been seen, had encouraged Milner to approach Zaghlul and had offered to act as intermediary, but had proved ineffectual. Now Milner and the Foreign Office persuaded themselves that Adli would deliver what Zaghlul had withheld. Allenby was instructed to advise Fuad to appoint Adli as prime minister. Fuad objected that Adli did not represent any real party, and parties in a Western sense were anyway non-existent in Egypt. But official British advice to the Sultan was tantamount to an order. Adli was installed in March 1921 as prime minister and in due course began his negotiations.

Zaghlul pre-empted him. He published a new set of demands which went much further than what Milner had already conceded. He returned to Egypt and began a fiery campaign of agitation against Adli, one consequence of which was a particularly bloody riot in Alexandria in which foreigners were murdered, their bodies burnt by the rioters, and their houses looted. Adli went to London with Zaghlul's threats and fulminations in his ears. He negotiated during the summer and autumn of 1921, but did not dare accept anything which might leave him open to attack by Zaghlul. He returned home having demonstrated to the British that he was a broken reed, unable to stand up to Zaghlul and impotent to effect anything.

The initiative now reverted to Allenby in Cairo, increasingly impatient and brusque. Following discussions with his advisers, he suggested breaking the deadlock by the Government unilaterally ending the protectorate, and not, as Milner and all other British negotiators had hitherto insisted, in exchange for a treaty. If the Government would do this he would be able to find an Egyptian willing to form a ministry. Allenby was able to promise this because he had been negotiating with the same group of Egyptian politicians to which Adli belonged, who naturally thought that it would be greatly to their advantage if it was they who obtained the complete abrogation of the protectorate with no kind of quid pro quo, the more so that Allenby had very conveniently spiked Zaghlul's guns by arresting and deporting him. Allenby and his advisers had somehow become committed to them, to such an extent that if the Government refused his plan, he could not, he declared, honourably remain in his post. To start with, the Government jibbed at the High Commissioner's ultimatum. They recalled him to London on which Allenby—the Bull, as he was familiarly known—descended in the middle of February 1922, having beforehand twice offered to resign if his scheme was turned down. He had two meetings with ministers on the morning and evening of 15 February, adamantly and stubbornly maintaining that either he had his way, or he would resign. Lloyd George climbed down, fearing that if Allenby resigned he would be able, as a peer, to mount a damaging attack in the House of Lords on the Government's mismanagement of Egypt, about which indeed plenty could be said. Allenby had his way, and the Declaration of 28 February 1922 recognized Egyptian independence subject to four unilaterally defined reservations relating to the safeguard of British strategic interests, imperial communications, the protection of foreigners, and the Sudan. Egyptian governments, however, felt under no obligation to recognize the reservations.

The draft of this declaration had contained a sentence to the effect that the British government 'will view with favour the creation of a Parliament with right to control the policy and administration of a constitutional, responsible government'. Though this was omitted from the final version, there is little doubt that it was part of the deal which Allenby had made with the Egyptian politicians before executing the *coup d'état* against his own Government. The principal figure in this group was an ex-minister, Sarwat Pasha, who now became Prime Minister. He was not the Sultan's choice, nor did Fuad like the idea of a constitution fettering his own powers, but he was unable to withstand Allenby's pressure exerted on behalf of Sarwat and his friends. It was, then, by a foreign fiat that constitutionalism was introduced into the independent kingdom of Egypt, as it now came to be called. When Milner was negotiating in London with Zaghlul and Adli, an Egyptian minister who was then in London, Osman Sirri, visited a member of the Milner Mission who reported to Milner what Sirri had to say. He pointed out that the British had 'assumed responsibility which we must not in common fairness to the bulk of the Egyptian people surrender'. An autonomous Egypt ruled by a parliament, he declared, would deliver Egypt 'into the hands of the dominant class, who would manipulate elections and purchase votes—the whole system of administration by baksheesh would start afresh and the fellah would undoubtedly be oppressed'.[6] In their General Conclusions, as has been seen, the Mission themselves recognized that a representative assembly, which they considered should ratify the Anglo-Egyptian treaty which they had in mind, could not, in any substantive sense, be representative, that it would in effect establish an oligarchy likely to show little regard for the interests of the people. It was such a system of government which Allenby pressed on a reluctant Fuad. Now that the British had declared Egypt independent, the King (*malik*)—the grander title which he now assumed—was naturally very loath to abandon the unfettered autocracy which his forebears had exercised, and he saw, with the clarity which the defence of his interests required, that a representative assembly would simply mean that autocracy would now pass from his hands to a handful of politicians, whose exercise of power would likewise be unfettered. It would be, in other words, a re-enactment of what took place when 'Urabi had roughly pushed Fuad's own brother Tawfiq aside in 1881–2, when Midhat had tried conclusions with Abd al-Hamid, when Mustafa al-Khaznadar had briefly supplanted the Bey as the real and effective ruler of Tunis. It is most doubtful if Allenby had the political sense to see this. In any case, before executing the *coup d'état* of February 1922 against his own Government, Allenby had engaged himself to Sarwat and his friends, and the engagement required that a constitution and a parliament be pressed on the unwilling Fuad.

Sarwat and his friends belonged, like Zaghlul and his followers, to the official classes: high officials, ministers, and landowning notables. Their intellectual antecedents lay in the pre-war group known as the People's Party which published a newspaper, *al-Jarida*, influential in educated, Westernized circles. Zaghlul too had gravitated to this group, which had been opposed to the Khedive's autocracy and had believed that Egypt's future ought to lie in constitutionalism, the rule of law, and representative government. The title of the party which Sarwat and his friends now formed, following the Declaration of 28 February 1922, well indicates their preferences and the ideals they professed. The party was called the Liberal Constitutional Party and its president was Adli, who at the first general meeting of the new party declared that a constitutional regime was the only form of government worthy of a country like Egypt which was steeped in civilization. Allenby, in a sense, was ironically the champion who did battle for this party against Fuad: autocrat pitted against autocrat.

Sarwat took office on 1 March 1922, on the express understanding that a constitution would be promulgated which would make provision for a parliament to which ministers would be responsible. A month later a commission composed of thirty members representing various interests and trends of opinion was set up and charged with producing a draft constitution. Its president was Rushdi Pasha, the prime minister who had resigned when the British rebuffed Zaghlul, and who had views similar to those of Adli and Sarwat, his erstwhile ministerial colleague. The Wafd would have nothing to do with this Constitutional Commission which Zaghlul denounced as the Malefactors' Commission. The draft which the Commission produced at the end of six months was, as might be expected, heavily influenced by European constitutional ideas and models. It declared that 'All authority derives from the nation'. Governmental powers were divided into legislative, executive, and judiciary. Legislative power was exercised by a two-chamber parliament in conjunction with the King, while executive power was to be exercised under the King by a council of ministers who would remain in office so long as they retained the legislature's confidence. The King had power to appoint and dismiss ministers and to dissolve the parliament. He also had a limited power to veto legislation. From a commentary accompanying the draft constitution which the Commission published, it is clear that they took for granted certain assumptions about the manner in which some of the crucial provisions would be applied. A gloss on the article empowering the King to appoint and dismiss ministers declared that it was the convention that the King would choose a prime minister and appoint other ministers on his advice. The assumption was that a prime minister would be chosen because he was supported by a majority in the parliament. Such a majority, the Commission also assumed, would naturally represent the wishes of the majority of the people as ascertained by means of elections. Hence, also, the right of dissolution of the parliament given to the King, enabling the nation to give its verdict if there was disagreement or deadlock between legislative and executive. What the Commission failed to consider was that elections—the mainspring of their scheme—might serve to register not the wishes of the electorate, but the designs of those who were in a position to manipulate elections, and for whom electors would vote because they were already in the seat of power. This was Cromer's fear when, in his last report before he retired in 1906, he warned that 'under the specious title of free institutions, the worst evils of personal government would reappear'. In patronizing representative government, Cromer's successor obviously had no fear of the outcome—possibly did not even know, soldier that he was, the meaning of such fear.

The powers which the Commission reserved for the King in their draft did not satisfy Fuad. Compared indeed to the limitless prerogatives with which his office had been endowed, the role of a constitutional monarch was limited and limiting. Of his prime minister, Sarwat, he disapproved greatly at any rate, and he now manœuvred to bring about his downfall. He let it be known that he desired Zaghlul's recall from exile, but that Sarwat—not surprisingly—was opposed to a policy of leniency. The Wafd increased its attacks on the administration, denouncing it as the creature of the British, and as having been imposed by martial law. Finally one of the King's men, Hasan Nash'at, organized a demonstration against Sarwat by Azhar students to take place on a forthcoming visit which the King, escorted by his ministers, would make to the mosque for Friday prayers. Sarwat heard of the plot and he preferred to resign at the beginning of December 1922, rather than be subjected to this humiliation. The resignation was accepted within the hour. Two ministries followed Sarwat in quick succession. During their tenure the draft constitution was amended in order considerably to increase the King's prerogatives. After Sarwat's resignation, rumours were rife that the clause asserting that the nation was the source of authority had been removed from the draft. When finally promulgated, the text was found to retain this particular clause; it was, however, accompanied by a curious and significant gloss, contributed by the Minister of Justice. 'The principle that the nation is the origin of all authority', this gloss affirmed, 'is not in contradiction with the principles of the Islamic, monarchical, and absolutist governments, because those monarchies used initially to depend on the explicit or implicit consent of the people represented by its elders and notables.' The democratic tenor of the clause is here ingeniously used to support an autocratic conclusion. The substantive changes effected in the draft gave the King power to confer civil and military rank, decorations, and titles at his own discretion. It also allowed him to appoint and dismiss at his discretion military officers and diplomats. Again, it increased the number of senators the King could appoint, and it put the presidency of the Senate absolutely in his gift. Religious endowments and Muslim religious institutions—of enormous importance financially and politically—were left solely under the King's control, until Parliament should legislate otherwise. The draft constitution had entrenched a number of articles to guard against their amendment or abrogation. In the final version, the entrenchment disappeared, to be replaced by a provision to the effect that no proposal was to be entertained which had for its object change in 'the provisions guaranteed by this Constitution concerning the representative form of government, the order of succession to the throne, and the principles of liberty and equality'. Except for the order of succession, which is precisely defined, and the protection of which is thus quite specific, the other provisions are very vaguely phrased, and what would constitute a change in them such as to be prohibited by this clause, remained very much a matter of opinion and therefore open to dispute.

The Constitution was promulgated on 19 April 1923, and in accordance with its provisions elections to a chamber of deputies had to be carried out. These took place in December 1923. In his struggle with Sarwat, Fuad had enlisted the help of Zaghlul and his Wafd. He now also had to make sure that the

Liberal Constitutionalists, whom he disliked for daring to attempt to limit his powers, should be trounced in the elections. There were many indications during the year preceding the elections that the King and the Wafd were, as they say, in cahoots. At one point, in the aftermath of Sarwat's resignation, there was a suggestion that Adli, now President of the Liberal Constitutionalists, might form a government. Wafdist incitement and terrorist incidents were stepped up in order to discourage him from such a venture. Hasan Nash'at was implicated in these activities—no doubt on Fuad's orders—and the British required his departure from Egypt, not allowing him to return until after the Constitution had been promulgated. Allenby repeatedly protested against these outrages, some of which involved the assassination of Englishmen, and as repeatedly was told by each of the two prime ministers who followed Sarwat between his resignation and the elections, that these incidents were the consequence of not paying heed to the demands of the majority, i.e. the Wafd. In May 1923 Allenby told his Government that Fuad's association with the Zaghlulists was deliberate and undisguised. Zaghlulist newspapers were receiving large subsidies from the Palace, and Palace influence was responsible for the recrudescence of Zaghlulist influence in the country—having previously receded with Zaghlul's second banishment and the formation of Sarwat's administration. Zaghlul returned to Egypt in September 1923 in order to fight the electoral battle, in which, it would seem, the Palace exerted its influence on his side. Hasan Nash'at took a message to Zaghlul that the King would be pleased if he became prime minister after the elections. Whether this was sincere may be doubted for, as he told a confidant at the time, he also believed that he could impose his own nominee. Zaghlul, however, swept in with a vast majority, and Fuad's complicated intrigue went awry. He had no alternative but to appoint as prime minister the people's tribune who claimed to be the head not of a party, but of the whole nation. In the year or so that he held office, Zaghlul sought to overawe the King by organizing noisy demonstrations against him. How far he would have gone in this course of intimidation is impossible to say. Zaghlul was undone by his imperfect control over his terrorist apparatus, the head of which, Abd al-Rahman Fahmi, was pardoned and released when he came to power. The terrorist apparatus now encompassed the assassination of the British Governor-General of the Sudan and head of the Egyptian army, Sir Lee Stack, and Zaghlul had to resign at the end of 1924.

From the very beginning, the workings of parliamentary government, judged by the standards set down in the Constitution, were seriously flawed. Even though Zaghlul had won the elections by an overwhelming majority, half of the members of his government were not members of the Wafd, but rather the King's nominees. The bargain which led to these appointments must have been struck before the elections when Fuad gave his support to the Zaghlulists and let their leader know that he would accept him as prime minister. When Stack was assassinated, Allenby imposed heavy penalties on Egypt, the most serious of which heavily reduced the Egyptian presence in the Sudan, which, following its re-conquest and the destruction of the Mahdist regime, had been governed, since 1899, by an Anglo-Egyptian condominium. It was one of Fuad's great ambitions to be recognized as King of both Egypt and the Sudan. He had insisted that this title should be included in the Constitution, but an ultimatum from Allenby forced him to desist. That the Egyptian position in the Sudan should be so diminished was a great blow to Zaghlul's prestige. This was one substantial reason why his position became untenable and why he had to resign. The King then dissolved the Parliament elected the previous year. A new administration was formed composed of King's men in coalition with some Liberal Constitutionalists, and new elections held. Probably because there had been no time to dismantle the Wafdist network in the country, the elections produced a Wafdist majority. This seems to have been the only occasion when elections produced results not in accordance with the wishes of whichever power happened to be preponderant in Cairo. The new Parliament was immediately dissolved and the administration did not find it necessary to call new elections. During the currency of this administration, the country was in effect ruled by the Palace.

Allenby's successor as High Commissioner, Lord Lloyd, considered this state of affairs undesirable since there was nothing to oppose Fuad's despotism and the abuses of power which naturally resulted from it. Under the Declaration of 28 February 1922, the British could not directly regulate Egyptian affairs, as they had done after 1882. Yet they had substantial interests which could be affected, and were sometimes seriously threatened, by misgovernment. Regardless, therefore, of the self-denying ordinance which the Declaration represented, they could not totally abstain from interfering in Egyptian politics. Their interference, however, had to be intermittent and to take place behind the scenes. It could, with reason, be denounced as illegitimate, and they could also be blamed by all factions, whether in or out of power, for whatever went wrong in Egypt. It is this situation which led Lloyd to write in 1927, when he had had some experience of his post and its predicaments, that 'We have magnitude without position; power without authority; responsibility without control.'

Lloyd, then, signified to the King that new elections were in order. New elections were held, which resulted in a Wafdist majority. The majority was Wafdist because the King had lost out in his bid for total power. The Wafd was still led by Zaghlul, but Lloyd vetoed Zaghlul as prime minister. The administration was a coalition of Liberal Constitutionalists and Wafdists under Adli and, following his resignation, under Sarwat, who resigned in March 1928, having failed to negotiate a treaty with Great Britain. Following the unilateral Declaration of 28 February 1922, an Anglo-Egyptian treaty had become the great issue for Egyptian politicians, who vied with one another in attempting to negotiate a treaty which would whittle down and finally eliminate British power in Egypt and the Sudan. In other words, the aim was to do away with the four subjects which the Declaration unilaterally reserved to Great Britain. Here, the maximum demands made by one political leader immediately and *ipso facto* became the minimum that any other leader dared to entertain. This ratchet-like mechanism profited Zaghlul greatly in 1919–21, and Allenby's impetuous initiatives ensured for it a long life—in fact until 1952, when its incautious use proved a potent factor in the ruin both of the Wafd and the monarchy.

When Sarwat failed to obtain a treaty which would not lay him open to attacks by his rivals, the Wafd, with its parliamentary majority, claimed the right to form a government. Zaghlul had died in 1927, and with him disappeared the reason for Lloyd's veto against a Wafdist prime minister. Mustafa al-Nahhas succeeded Zaghlul as head of the Wafd and Sarwat as prime minister. But the King, who disliked both the Wafd and Nahhas, was able shortly afterwards to dismiss him. It was disclosed that the Prime Minister and two of his parliamentary colleagues had contracted, for a fee of £130,000, to take steps to return the substantial property of a member of the royal family, Prince Seif al-Din, which had been sequestered by Fuad on the alleged grounds of his relative's derangement, to Seif al-Din's mother. The ground of dismissal was that this contract showed improper and corrupt use by Nahhas and his friends of their political influence and parliamentary position. Nahhas was succeeded by Muhammad Mahmud, now President of the Liberal Constitutionalists. He had previously served in the coalition government which succeeded Zaghlul at the end of 1924. He had also served in Nahhas's government which included some Liberal Constitutionalists, and had resigned from it shortly before the prime minister's dismissal. Muhammad Mahmud formed a government which included both Liberal Constitutionalists and King's men. Parliament was dissolved, elections postponed for three years, and Muhammad Mahmud ruled by decree.

Muhammad Mahmud's political record through this period exhibits in a nutshell the leading characteristics of Egyptian political life following Allenby's coup of February 1922. The son of an important landowner and educated at Balliol College, Oxford, Muhammad Mahmud had been, from its inception, a leading member of the Liberal Constitutionalist Party, which proclaimed its attachment to constitutionalism, the rule of law and representative government. The Party was also opposed to Zaghlul, and what it denounced as his demagogy and unrestrained appetite for power. Yet Muhammad Mahmud found it possible to serve in 1925 in an administration which dispensed with elections and a parliament, and in 1928 to accept office as a colleague of Wafdist ministers, even though change of leadership had made no difference to the nature of the Wafd. Now, as the head of a government, this Liberal Constitutionalist was happy himself to dispense with elections and a parliament. His later record was no different from the earlier, or from that of the great majority of his party colleagues. In spite of its principles and its programme, then, the Liberal Constitutionalist Party was simply an opportunistic faction seeking the fruits of office and power, willing in pursuit of them to perform any political somersault, and work with anyone, whether King or Wafd, of whom they had been supposed to disapprove. Other political parties were likewise factions solely dedicated to the pursuit of power and profit: the ephemeral Unionist Party formed of King's men following Zaghlul's downfall in 1924; the equally ephemeral People's Party formed by Ismail Sidqi when he was prime minister between 1930 and 1933; the Sa'dists who quarrelled with Nahhas and seceded from the Wafd at the end of 1937; the Independent Wafdist Group (*al-kutla al-wafdiyya al-mustaqilla*) formed by William Makram Obayd, who had long been Nahhas's right-hand man and who broke with him in 1942; and the National Party, established by Mustafa Kamil in 1907, who had begun his anti-British and nationalist activities under the aegis of Khedive Abbas, whose programme and slogan of complete independence for Egypt was later appropriated by Zaghlul, from whose ranks Nahhas emerged, and which after 1919 led a dim and insignificant existence, supplying a minister or two to various cabinets in the 1940s.

The Wafd claimed to be different from all these parties. It claimed to be the voice of the Egyptian people, the only organization which could legitimately speak on their behalf. Such a claim, described in this way, has little substance. The Wafd was different from the factions described above owing to Zaghlul's tactics and his legacy. From 1919 onwards Zaghlul manifested great demagogic abilities. He could rouse by his oratory, and elicit the enthusiasm and sometimes the frenzy of the crowd. Genuine political representation is different. It involves a continuous dialogue between citizens and their political leaders—a dialogue which derives its substance and its significance from the variety of interests and opinions which find articulation within an orderly framework of constitutional and responsible government. This did not exist in Egypt, and the Wafd, far from being representative, was simply yet another faction within the official class. It was, however, a faction whose leader had discovered that he was able to mobilize the mass which had been discomforted by the effects of war, and had then converted his discovery into a system. Zaghlul thus made his faction into a movement, something inchoate, restless and turbulent, and at times fanatical, which he could manipulate, when opportunity offered, as a weapon against rivals and enemies. But for all that, the Wafd leaders were no more than participants in the factional politics which were the rule in Egypt under Fuad, and his son and successor Faruq. In this kind of politics, the policies of the British High Commissioner, and later Ambassador, buttressed by the army of occupation, the extensive influence of the royal palace, as well as the manœuvres of the various factions, all dictated which politicians were to enjoy, for the time being, the fruits and profits of power. In this respect the Wafdist politicians, beginning with Zaghlul, were no different from their fellow-politicians. When in power they looked upon the state and its administrative apparatus as a bowl of soup from which to sup themselves, and to feed followers and clients. Hence, for instance, at any change of government there would be a general post in which the officials appointed by the retiring administration would be ousted, and the friends of the new masters would come in. As has been mentioned, William Makram Obayd broke with Nahhas in 1942. For many years Secretary-General of the Wafd, when the break came he was the Minister of Finance in the administration of which Nahhas was the head. Following the break, which had been encouraged by the King, he compiled and circulated a Black Book which detailed the corruption, the peculations, and the abuses of which Nahhas and his entourage were guilty. By reason of his long service and eminent position in the Wafd, Obayd was very well informed about the activities of associates whom he had intimately frequented for some two decades. His Black Book holds up a mirror, no doubt distorted by spite and vengefulness, to the political class of Egypt under the constitutional monarchy. Had anyone else associated with another party or faction chosen to emulate Obayd, his revelations would have painted a similar picture.

The interactions between the policies of the King, the parties, and the British which governed political changes in Egypt after 1923 account for the fall of Muhammad Mahmud's Government in the latter half of 1929. In that year, a Labour Government took office in Britain. It, too, wished to conclude a treaty with Egypt which would establish Anglo-Egyptian relations on a less precarious footing than that offered by Allenby's unilateral Declaration of February 1922. It began negotiations with Muhammad Mahmud. Just as the earlier negotiations with Sarwat had failed, leading to his downfall, so now, too, negotiations with Muhammad Mahmud failed, and for the same reason, namely that what the British offered did not satisfy Egyptian demands, which party rivalries pitched at a maximum, and less than which no party, at the risk of fierce denunciation by its rivals, dared to accept. The British Government believed that the negotiations failed because they had been conducted with ministries which did not represent the majority of the Egyptians. The only party, they were persuaded, which did represent the majority was the Wafd, therefore the only party with whom negotiations could be successful. Muhammad Mahmud resigned and the King was obliged to call upon Nahhas to succeed him. He took office on 1 January 1930. New elections were held which naturally resulted in a Wafdist majority. The negotiations with Britain, however, failed and the King dismissed Nahhas in June and called upon Ismail Sidqi to succeed Nahhas. Sidqi dissolved the recently elected parliament and replaced the Constitution of 1923 with one which gave the King even greater powers: ministers were now responsible to him, not to the legislature; he initiated legislation; and he could veto legislation which had been passed by the Parliament. A new electoral law was also promulgated which replaced direct by indirect elections. Sidqi was dismissed in September

1933, and an administration formed of King's men succeeded him. It was followed by another similar one in November 1934. The King repealed Sidqi's Constitution and dissolved the parliament which Sidqi had had elected. Fuad now governed by royal decree without a parliament.

A complicated set of events in 1935–6, chiefly the renewed desire of the British Government to have a treaty with Egypt, which it continued to believe could only be negotiated with the Wafd, led to a revival in the fortunes of this party. Fuad had to reinstate the 1923 Constitution and electoral law. New elections were held in May 1936, the Wafd being assured of the lion's share of seats. It now formed a government. It had agreed in advance to take part in a delegation representing the various parties which would negotiate an Anglo-Egyptian treaty.

A treaty was signed in August 1936. It seemed to have the approval of the Wafd, as well as the other parties. It provided for British troops to continue to be stationed in Egypt, and for the Sudan to continue to be governed under the condominium regime set up in 1899—both issues which Egyptian negotiators had previously refused to concede. In the meantime Fuad had died. He was succeeded by his son Faruq, who attained his majority in July 1937. Like the father, the son had an invincible dislike of the Wafd, and mistrust of the ambitions of its leader. Very soon sharp disagreements arose between them. Nahhas was dismissed at the end of December 1937. Muhammad Mahmud, the Liberal Constitutional, succeeded him. He dissolved the Parliament with its overwhelming Wafdist majority and called new elections. These took place in April 1938 and ratified the King's dominance: only twelve Wafdists were elected to the new parliament.

This Parliament lasted until February 1942. It witnessed a succession of governments made up of anti-Wafdist factions, which came into being and then disappeared in response to Faruq's caprice, or else to the various contingencies which the Second World War occasioned. As in the First World War, Egypt served as a British base, the loss of which, after the fall of France and the entry of Italy on the side of Germany, would be extremely serious. The reverses which Britain sustained in

1940–1, and the presence in large numbers of Italian and German troops in Libya, increased pro-Axis sympathies in Egypt on the part of many ministers, high officials, and Faruq himself. The British judged that they had to secure public tranquillity in Egypt and to ensure that their forces and installations were not at the mercy of Axis sympathizers, at a time when Axis forces, commanded by General Rommel, were threatening Egypt from the Western Desert. To avert an internal threat to the rear of the Allied forces, they determined to impose a Wafd government on Faruq, in the expectation that the Wafd, which they considered to represent the majority, which was opposed to the King and supposedly hence to the King's pro-Axis proclivities, and which had agreed to the 1936 treaty, would be willing and able to keep Egypt quiet and loyal to the Allied cause. On 4 February 1942 the British Ambassador, therefore, surrounded the royal palace with troops and tanks, and confronted Faruq with an ultimatum: either to appoint Nahhas as prime minister or to abdicate. Nahhas was appointed. The episode was replete with ironies. The Wafd which, since Zaghlul, had claimed to be in the vanguard of the anti-British struggle, was now installed in power through a British *coup d'état*. The British, again, who had believed that the Wafd represented the majority of the Egyptian people, and that an agreement made with it would enable them to preserve their interests without intervention in Egyptian politics—a mirage which had beguiled Milner, now had to resort to forcible means in order to bring into office those representatives of the majority. The Foreign Secretary and the Ambassador to Cairo, who respectively approved and carried out the *coup d'état*, were the very same who, in pursuit of the mirage, had, with great satisfaction and self-congratulation, negotiated the treaty a mere six years before.

Nahhas dissolved the 1938 Parliament, and the elections which followed—boycotted by the anti-Wafdists—produced an overwhelming Wafdist majority. In October 1944, finding that the British no longer insisted that Nahhas should remain in power, Faruq dismissed him. The Wafdist Parliament was dissolved, and the elections which followed—which the Wafd again boycotted—produced in January 1945, a wholy anti-Wafdist parliament. The Wafdists were succeeded by a government drawn from anti-Wafdist factions, headed by Ahmad Mahir, who had been a Wafdist, but had seceded with a few colleagues in 1937 to form the Sa'dist Party. In February 1945 he was murdered for having moved the Parliament to approve a declaration of war against the Axis, in order to enable Egypt to become a member of the newly-founded United Nations. The murder heralded and indicated a significant change in the climate of Egyptian politics. Ahmad Mahir's murderer was believed to be a member of the Society of the Muslim Brethren. Founded in 1928, the Society aimed originally to bring aid and comfort to the increasing multitudes of immigrants flooding into the cities, who lived in material poverty and spiritual disorientation. It attained great success in its objects, and by the end of the 1930s it had a very large number of followers and administered many enterprises which provided help, and organized self-help, for its members. This great success inspired in its founder, Hasan al-Banna, political ambition. The aim of the Brethren became to institute a purely Islamic polity in Egypt. In the late 1930s it took up the cause of fellow-Muslims who were struggling against the British mandate and the Zionist settlement in Palestine, at a time when interest in the Palestine problem was not very widespread within the official and intellectual class. Owing to his large following, Banna began to be courted by various political leaders and simultaneously to be feared by them. Banna did not take part in the political game as Egyptian leaders played it between the wars. His methods were remarkably similar to those which had brought so much initial success to Zaghlul. On the one hand the Brethren preached to, and mobilized the mass in order to bring pressure on the government, and, on the other, Banna organized a secret apparatus trained and armed to carry out terrorist operations. To use the distinction made above, the Muslim Brethren aspired to be not a political faction, but rather a political movement.

They were not the only ones to do so. A Young Egypt Party, founded in 1933, followed the example of the Fascists and the

Nazis in Europe in attempting to gain power by mobilizing supporters for a variety of extra-constitutional actions. Thus Young Egypt tried to organize its adherents into a body of Greenshirts, who sought to intimidate opponents and the government by violent action on the streets. In response the Wafd organized a body of Blueshirts, and the two groups often engaged in street brawls and clashes with one another. These extra-constitutional forms of political action became very attractive to the young in a society which did not have the wherewithal to satisfy their expectations. These expectations were aroused by the exposure of increasing numbers of children and young men to a European-style education in schools and universities. But the expectations could not possibly be fulfilled. Egypt was a poor country, dependent on agriculture, and in particular on cotton, its main export, which was subject to the—sometimes disastrous—vagaries of a world market on which its price was dependent. Agricultural land was strictly limited by the availability of water, and water came from only one source, the Nile. This rather poor country had to cope with an inexorable and cumulative increase in population: from being eleven and a quarter million in 1907, it had become almost sixteen million in 1937. The Egyptian economy was unable to provide work which might secure a tolerable livelihood to these millions.

This increase in population was much more noticeable, and its consequences much more acute, in the cities which attracted migrants from the countryside. The increase in urban population was therefore appreciably greater than in the country as a whole. In 1907 Cairo had 678,000 inhabitants, in 1937 1,312,000. Housing, public transport, health, and municipal services lamentably failed to keep pace with inexorably increasing multitudes for whom, anyway, little employment was available.

The expectations of students and graduates, doomed to disappointment as they were destined to be, were further exacerbated by the rhetoric produced in the course of the ideological struggle before, during, and after the Second World War. For instance, Nasser and Sadat, successively Presidents of Egypt after the military *coup d'état* of July 1952, together with many other officers who took part in the *coup*, became greatly attracted as schoolboys in the 1930s to the radical ideologies of Young Egypt and the Muslim Brethren. These ideologies inspired in them visionary hopes of a total change in the political and social condition of Egypt. In his well-known *Philosophy of the Revolution*, Nasser (who had actually belonged to Young Egypt) expresses in an arresting manner the dreams of these schoolboys, which remained with them through their early manhood, impelling them to a style of political action both conspiratorial and violent: 'Our life was, during this period, like an exciting detective story. We had great secrets; we had symbols; we hid in the darkness and arranged our pistols and bombs side by side.' Nasser and his friends yearned for 'positive action'. In his inflamed mind, he tells us, political assassinations 'blazed' as the only positive action 'from which we could not escape, if we were to save the future of our country'. In his opuscule, Nasser goes on to describe an attempt to assassinate an Egyptian public figure. Following the attempt, he escapes from the scene, goes home, and lies on his bed, a prey to tumultuous and exalted feelings. He would say to himself: 'What is important is that someone should come who should come.' The messianic tone is unmistakable: he who should come should come.

It is at any rate clear that neither the assassinated Ahmad Mahir, nor his fellow Saʿdist Mahmud Fahmi al-Nuqrashi, who succeeded him and was in turn assassinated by the Brethren in 1949, was he 'who should come'. It is also ironical that these two particular figures should have been the victims of terrorists driven by a radical vision, for they themselves, as Wafdists, had been implicated in the earlier terrorism which had issued in Sir Lee Stack's assassination in 1924.

The political climate confronting Nuqrashi's administration in 1945–6 was tense and volatile. When the war ended the restraints imposed by martial law and press censorship were lifted, and various groups and factions felt free to agitate against the government in power and to air a variety of social and political grievances. These grievances found expression ostensibly in agitation against the Anglo-Egyptian treaty. The treaty, which had a twenty-year currency, allowed the stationing of British troops in Egypt, and retained the status of the Sudan as an Anglo-Egyptian condominium. In self-preservation, any Egyptian government was bound to demand the termination or at any rate the revision of the treaty. Nuqrashi eventually did so at the end of 1945, and the Labour Government in Britain declared its willingness to enter into negotiations. This exchange in turn triggered a vociferous agitation. A student demonstration on 9 February 1946 which demanded the full and immediate evacuation of British troops clashed with the police, and there were dead and wounded. Some of the ministers in Nuqrashi's coalition thereupon resigned, and the Prime Minister decided that he could not carry on.

Sidqi followed him with a coalition of Liberal Constitutionalists and Independents. He too attempted to negotiate with Great Britain, but the negotiations came to nothing; the issue of the Sudan—of the Unity of the Nile Valley as it was called in official Egyptian discourse—was again the stumbling block. Like his predecessor, Sidqi was confronted with demonstrations and riots, in which the rioters themselves used arms against the police. It was clear that the deterioration of public order was worsening and that nothing the Government did could retrieve the situation. Sidqi resigned in December 1946, and Nuqrashi again took office with a coalition of Sa'dists and Liberal Constitutionalists. The new Government was unable to prevent public disorders or political assassinations, for which latter the Muslim Brethren were believed to be responsible.

In November 1947 came a new development which was to have far-reaching effects on Egypt under the monarchy, as well as for decades after the deposition of Faruq and the institution of a republic. In that month, the United Nations voted in favour of the partition of Palestine into a Jewish and an Arab state. Ever since the middle of the Second World War, in a radical departure from previous attitudes and policies, Egypt had embraced pan-Arabism and had become a founder-member of the League of Arab States established in 1945. Negotiations which culminated in the foundation of the League began in 1943, when Nahhas was in power. Both he and Faruq saw possibilities of aggrandizement in this new pan-Arab policy. Faruq, in particular, hoped that, better than the caliphate, it would make him, as the ruler of the most powerful state in the League, the unchallenged leader of the whole Arab world. Palestine was the main issue which the League had to deal with from its foundation. Its members took up the cause of the Palestine Arabs and voiced absolute opposition to Zionism and to Jewish immigration in Palestine. When the partition was approved at the United Nations, the Arab League states declared that they would adamantly oppose its implementation. How they were to do so, however, remained uncertain almost until the end of the British Mandate in May 1948. It was, in fact, only at the last minute that the Arab states decided to send their own armies and prevent partition by force. Nuqrashi's Government, in particular, was very reluctant to send the Egyptian army to fight the Zionists and it was not until the end of April that, on Faruq's insistence, the army received orders to prepare for the invasion of Palestine. It was, in all respects, an inauspicious venture. The army was not prepared for a campaign outside its borders, and when it was defeated, discontent among the officers became very great and was a material cause of the military *coup d'état* of July 1952.

The tension which was increasingly manifest in Egypt since the end of the World War continued to increase, fomented by various radical groups. Of these, the most formidable were the Muslim Brethren. The war in Palestine enabled the Government to declare martial law, and this in turn allowed it to dissolve the Society of the Brethren, to confiscate its assets and prohibit its activities. This was done on 8 December 1948. On 28 December, a member of the Brethren, as has been said, assassinated Nuqrashi. On 12 February following, the Supreme Guide of the Brethren, Hasan al-Banna, was in his turn assassinated.

A fellow Saʿdist, Ibrahim Abd al-Hadi, succeeded Nuqrashi, heading a coalition of non-Wafdist ministers. The tension which had been manifest since 1945 increased as a result of the fiasco in Palestine, and the budgetary difficulties which the war entailed. Though martial law continued in force, and though the Brethren and other radical groups had been banned and many of their members interned, yet clandestine agitation against the government and the regime continued. The Wafd had been out of power since 1944 and no doubt helping to foment the agitation, while Faruq and his advisers were increasingly disquieted by the accumulating economic, social, and security problems besetting the regime. The Wafd and the Palace eventually came to an agreement, and a coalition headed by an Independent in which the Wafd participated, was formed in July 1949. The entrance of the Wafd into the Government showed that they were now in the ascendant. Faruq had been forced to ask for their co-operation, and the January 1950 election ratified the victory of the Wafd over the King. Nahhas then formed what was to be his last administration.

The main issue before the new Government was again the modification of the Anglo-Egyptian treaty in respect of the Sudan, and the stationing of British troops in Egypt. Anglo-Egyptian negotiations went on from March 1950 to October 1951. They ended in deadlock. The Government thereupon decided on the unilateral abrogation of the treaty as well as of the condominium agreement of 1899. It had, however, no means of dislodging the British forces from their base on the Suez Canal, the presence of which it now deemed illegal. The issue was mainly left to so-called guerrillas acting on their own, and composed of the same radical elements which had been trying in various ways to destabilize Egyptian governments from 1946 onwards. These elements had been, to some extent, restrained by the operation of martial law from May 1948 onwards, which the Wafd, however, had proceeded to repeal in May 1950. Since the guerrillas were carrying on what the Government declared to be a national struggle, it was very difficult for it to put a brake on, or hinder, their activities. On the contrary, such patriotic activities benefited from the sympathy and help of various officials and administrative bodies. The attacks on the British base were, however, not allowed to go unanswered, and numerous clashes and punitive actions took place. The climax came on 25 January 1952, when the

British command decided to expel the Egyptian police and gendarmerie from Ismailiyya in order to deny the city to elements attacking its troops and installations. On order of the Minister of the Interior, the Egyptian police commander refused to comply with the British demand. British forces thereupon destroyed the police barracks, killing in the process over fifty policemen. The following day, 26 January, the mobs burned Cairo. Hundreds of buildings were destroyed, and many foreigners killed and wounded. Martial law was quickly proclaimed, and the Egyptian army ordered to clear the streets. On 27 January the King dismissed Nahhas and his Government.

This was, in effect, the end of the Wafd as well as of the monarchy. On 23 July 1952 a military *coup d'état* took place. On 26 July Faruq abdicated and left Egypt never to return, and the dynasty established by his great-great-grandfather came to an end. With the dynasty went the Constitution of 1923, abolished in December 1952, and the political parties, dissolved, banned, and deprived of all their assets in January 1953. The Wafd, which had claimed to represent the Egyptian people, and with whom alone, successive British governments were persuaded, a treaty could be negotiated, from one day to the next disappeared without trace. The thirty years of constitutional monarchy had meant in fact misgovernment, arbitrariness, and corruption. To judge by their actions, none of the political parties believed in, or had the desire, to make constitutional government workable. Its last years, after 1945, saw increasing tension and disorder. If, during the following decades, people came to regret its passing, this was surely because if Fuad and Faruq, Zaghlul, and Nahhas chastised the Egyptians with whips, their successors after the downfall of the monarchy were to chastise them with scorpions.

2. Parliamentary Government in the Kingdom of Iraq

As we have seen, constitutional government was introduced in Egypt by the fiat of the British High Commissioner. He was able to do so because Egypt was, after 1882, under the control of the British army of occupation. The British were, at about the same time, able to introduce a constitutional government in three former Ottoman provinces, Basra, Baghdad, and Mosul, because they conquered them in the course of the First World War, which the Young Turks had rashly entered on the side of the Central Powers in 1914. These three provinces were of particular strategic and economic interest to Great Britain, and in the settlement which followed they were given a 'mandate' to administer these provinces on behalf of the newly-established League of Nations.

According to the Covenant of the League, some countries which had formed part of the territories of the Central Powers or had been governed by them, and which were not yet ready for independence, were to be prepared for self-government by a Power mandated by the League to this end. In the case of the three ex-Ottoman provinces, Great Britain was designated as the Mandatory Power. These three provinces were to become a new state—Iraq—whose governmental institutions were to be set up and supervised by Great Britain until such time as the new state became fit to enjoy full sovereignty. Shortly after Great Britain was designated as the Mandatory Power, it contrived that the provisional government it had set up should invite Faysal, third son of Husayn, Sherif of Mecca who in 1916 became King of the Hijaz, to be the ruler of the new state. A plebiscite, similarly contrived, endorsed the invitation. Faysal was proclaimed King of the new state in July 1921. Its government, the Council of State affirmed, would be 'constitutional, representative, and democratic . . . limited by law'.

Faysal was a foreigner in the kingdom he was brought to rule. His father had, following secret and tortuous negotiations with the British in Cairo, raised the standard of rebellion against his Ottoman suzerains in July 1916. Faysal eventually became the commander of a so-called Northern Arab Army, financed and armed by the British. Colonel T. E. Lawrence, known as Lawrence of Arabia, was attached to Faysal, and became his fervent partisan. At the very end of the war, Allenby, who commanded the forces fighting the Ottomans in the Levant, forbade his troops to enter Damascus, in order that Faysal and his followers should claim its conquest. This was done in order to withdraw Syria from the French zone of influence in which it had been included by the Anglo-French Agreement (known as the Sykes–Picot Agreement) of May 1916. Faysal was enabled to establish his own administration in Syria, against strenuous French objections. In the end, the French were able to assert their rights under the Agreement, and the British left Faysal to make whatever arrangements he could with them. In this he was unsuccessful, his hot-headed followers in Damascus insisting on resisting French demands, by force if necessary. It soon came to a clash, and in July 1920 the French evicted Faysal from Damascus, where his followers had proclaimed him King of Syria the previous March.

Faysal, however, had powerful friends in London, in particular Colonel Lawrence, now a civil servant in the Colonial Office, and they pressed for him to be made King of Iraq. Their efforts were successful because that same summer which saw Faysal evicted from Damascus also witnessed a tribal uprising in the Euphrates, south of Baghdad. The uprising was a predominantly Shiite affair. It had been fuelled by the discontent of some powerful tribal leaders because rivals of theirs had been favoured by the British. The ambitions of Shiite divines in the Shiite shrine cities of Najaf, Karbala, and Kazimayn, dreaming of political power in a country predominantly Shiite, also had a large part to play in sparking the disturbances, since Shiite divines traditionally enjoyed very great respect and prestige among the devout mass of the Shiite tribesmen. Before Faysal's downfall in Syria, agents of his regime also played their part in fomenting anti-British activities in the neighbouring territory, notwithstanding that Faysal's regime had been set up, and was armed, and financed by, the British. These officials and officers who from Syria organized attacks against the British in Iraq originated predominantly from the three Iraqi provinces, and they did not see why only Syria should enjoy an independence which was denied to them. All these groups contributed to the effervescence which eventuated in the uprising. What also contributed was the fact that the British authorities themselves seemed to pursue no coherent policy. When the area had been occupied, it was widely assumed that it would eventually be governed by, or under the supervision of, the Government of India. But from 1917 onwards, a variety of proposals and pronouncements propounding or promising 'self-determination' had followed upon one another. Their language and significance was quite unfamiliar to those for whom they were destined. They created confusion and raised grandiose but inchoate expectations. Such prospects greatly increased the temptation to force, by violent means, the hand of a foreign administration which had not yet had time to establish unquestioned and unchallenged title to obedience; an administration, furthermore, which showed itself hesitant in dealing with aggressive incursions mounted by Faysal's followers from Syria.

What made the uprising possible was that the tribes of the Euphrates were well armed, having picked up large numbers of weapons and much ammunition from the battlefields where British and Ottoman forces had fought. Thus, when the disturbances of 1920 ended, the British authorities were able to collect over 63,000 rifles and nearly three million rounds of ammunition. Arms in such quantities had never been seen before in the Euphrates. Also, the British military command seems not to have suspected the dangerous instability affecting the area, and when the rebellion broke out, they were taken unawares.

The disturbances lasted about three months. The British forces, having recovered from their initial unpreparedness, had little trouble in snuffing these out. However, their political consequences were as far-reaching as they were unexpected by the tribal leaders and the Shiite divines who had fomented them. The uprising gave Lawrence and his friends an opening to discredit the civil administration in Baghdad headed by Colonel A. T. Wilson, who was sceptical of the feasibility of self-determination in Iraq. Lawrence and his many influential friends in London proclaimed that Wilson was a tyrant who tried to put down a national revolt by people rightly struggling to be free. Their just grievances had to be satisfied, by setting up a national government to be headed by Faysal who, they claimed, was unanimously wanted by the people. The British

Government found the argument persuasive since it was under pressure to effect economies and it welcomed a settlement which would drastically cut down military expenditure in Iraq.

So it was then that Faysal found himself the King of a kingdom whose inhabitants had been, shortly before, unaware of his very existence. This new kingdom was heterogeneous in the extreme. The majority was made up of Shiites, in the mass semi-nomadic and illiterate tribesmen whose leaders, lay and clerical, had challenged British arms and had been decisively defeated. When they began their uprising, tempted by the blandishments of Faysal's followers, they had proclaimed that they desired him for a king. They no doubt thought that he would be only a figure-head, while they themselves would enjoy the substance of power. But now they were defeated and disarmed and Faysal had come with his own Sunni followers, for the most part ex-Ottoman officers imbued with the same antipathy towards the Shiites as the Ottomans they had served before deserting during the war to the Sherif of Mecca. Half the population or more was Shiite, a mass, as has been said, of semi-settled tribesmen concentrated in the south of the country, whose loyalty went to their tribal leaders and to the divines of the shrine cities, both of whom were now bitter and disaffected. They would prove somewhat difficult to govern. The Kurds, in the north-east, mostly tribal, constituted another fifth of the population. They also were difficult to govern and their chiefs resented rule by an upstart Arab regime from Baghdad, which they had never before regarded as the centre of authority. Any loyalty which they felt towards a state had gone to the Ottomans who had been their masters for centuries. What increased Kurdish leaders' sense of grievance was that they had glimpsed for a short moment the possibility of autonomy and even independence—a prospect held out to them in the Treaty of Sèvres between the allies and the defeated Ottoman government. The Treaty was signed in 1920, but owing to Mustafa Kemal's victory, it proved abortive. As for the Sunni Arabs who thus constituted a minority in the new kingdom, large numbers of them were nomads or semi-nomads who had little say in politics, and were utterly unable to participate in the

Western-style political institutions with which the Mandatory Power was equipping the new kingdom.

What, then, constituted the *pays politique* in Iraq was a very small group of ex-Ottoman Sunni officers and officials. Led by Faysal, who had his own position to establish, this group was now charged with establishing and strengthens a government which, as the Council of State had declared, would be constitutional, representative, and limited by law. It was, as the sequel proved, extremely improbable that these desiderata would come anywhere near realization. However, a Constitution was drafted and approved by a Constituent Assembly in 1924 which provided for a Chamber of Deputies, indirectly elected, and for a Senate wholly appointed by the King. The Constitution made provision for a Cabinet responsible to the Chamber of Deputies, and for a system of Courts which 'will be free from interference in their affairs'. The Constitution also stipulated equality for all citizens regardless of differences in nationality, religion, and speech. Personal freedom was guaranteed, as were the rights of property. Torture and exile outside the frontiers were forbidden. It was indeed a model constitution, but a model which reality was never in the least able to emulate.

The Kingdom which the British established and which Faysal ruled was in fact highly centralized and governed by a very small group of Sunni politicians. It was, as has been seen, extremely heterogeneous in point of religion, ethnicity, and language, and the nomadic and semi-nomadic character of much of its population made for frequent turbulence and disorder. However, the Shiites had been broken by the British in 1920, and after the departure of their troops the British maintained airbases in the country sufficient to quell any incipient rebellion in the south. In the north, too, Kurdish chieftains tempted to assert themselves against Baghdad were kept in order by aerial bombing or by the threat of it. The presence of the British Royal Air Force, advice and help of British officials, British diplomatic support which defeated Turkish attempts in the 1920s to annex the ex-Ottoman province of Mosul where Turkey had many Kurdish and Turkish sympathizers—all of this made it possible for the new rulers to establish a meticulous country-wide network of administrative control with which to overawe and suppress opposition, which nevertheless kept on erupting. A British Air Force intelligence officer stationed in the provinces describes, in a remarkable passage, the style and methods of a rule which knew no checks and balances and which now flourished under the umbrella of British protection:

Here the structure of government is shaky and impermanent. Moreover, such control as government exerts over one's affairs is a terribly personal one. Government is not, as with us, a machine which grinds out laws; takes money out of one's pocket or puts money into it; forbids one to do this and permits one to do that with dispassionate implacability. It enters into the house here. It knows that you have four sons and that one of them is a post office official in Mosul. It knows that you have Turkish leanings, and that, as a natural consequence of such, you are not to be trusted. It knows that you were friends with Hamid Khuluf before his exile, that you are, therefore, probably sending information to Persia and that it must on that account consider in a fresh light what to do with your claim for water rights against Muhammad Derwish. It makes a vital difference to the issue of this or that land case whether Abdul Qadir happens to be Mutesarif at the time of its coming before the courts or whether he has been transferred to another district and someone else is sitting in his place . . . It is this grossly personal element in the all-pervading activities of government which evokes from the uneducated people that quality which we are all too apt to dismiss as insincerity, but which is, in reality, nothing but the inevitable compromise of any simple man chased by the bogey of insecurity . . . He must propitiate and speak fair words. His position is unstable. There is no permanence. He knows that the fact as to whether the official has a good or bad opinion of him will affect his private life vitally. He feels the ground shifting beneath his feet. It is the same with the official himself when he addresses a superior. He too feels the ground quaking beneath him, feels his confidence welling out. He may be sacked because his enemies have spoken ill of him. There will be no redress for him, no rehabilitation, unless he has influence in high places.[7]

Relations between rulers and ruled, as depicted here, necessarily precluded liveliness or vigour in the constitutional politics which the new kingdom was supposed to follow. In fact, all through the life of the monarchy, from the first parliament elected in 1925 to the sixteenth and last, elected in 1958, it was the government which decided who was to be elected, and it was the responsibility of the provincial governors to ensure that those favoured by the authorities were returned. For example, in the last general election under the monarchy, that of May 1958, in 118 constituencies out of 148, only one candidate offered himself and even the formalities of an election could thus be dispensed with. In effect members of parliament were appointees of the government in power.

Even though parliament was of no consequence, and governments in control of the administrative apparatus did not feel that they had to render account to the people's representatives, yet, as Macdonald remarked in the passage quoted above, the structure of government was shaky and impermanent. Such a condition is explained by the fact that those who ruled Iraq under the monarchy were a very small group who were continually jockeying with one another for power, and the success or failure of whose intrigues and plots meant for their dependants and clients the difference between prosperity and comparative impoverishment. Thus, between the proclamation of Faysal as King on 23 August 1921 and the murder of his grandson Faysal II on 14 July 1958, no less than fifty-eight cabinets succeeded one another. Some 175 ministers served in these short-lived cabinets, the majority of whom were ephemeral figures of no political weight or significance, either rewarded with office for some service, or replacing someone fallen from favour, or else used for some passing *combinazione*, and who, on leaving office, fell back into the obscurity from which they had briefly and fortuitously emerged. The real masters of the Iraqi political game—such as it was—were no more than a dozen or so Sunni Arabs: men such as Yasin al-Hashimi, Nuri al-Sa'id, Ja'far al-Askari, and Jamil al-Midfa'i, ex-Ottoman officers who had thrown in their lot with Faysal; or Rashid Āli al-Gaylani, Hikmat Suleiman, Tawfiq al-Suwaydi and his brother Naji, mostly ex-Ottoman officials whom Faysal needed to man his administration. Such men were cunning and ruthless enough to rise to the top and to become leaders of factions. Nuri al Saʿid became prime minister fourteen times, Jamil al-Midfaʿi seven times, Rashid Āli four times, Tawfiq al-Suwaydi three times, and Yasin al-Hashimi twice.

In 1932 Iraq became a fully independent state. Great Britain, the Mandatory Power, had persuaded the League of Nations that Iraq was now fully fit to enjoy complete sovereignty. In advocating the ending of the mandate, the British High Commissioner in Baghdad strove to banish the doubts and anxieties of the League's Permanent Mandates Commission by a fluent analysis and a stout show of confidence:

Nobody would think of excluding a Moth aeroplane from an international exhibition merely because it is not so swift as (say) a three-engined Fokker . . . Similarly, I submit it would not be right to attempt to argue that Iraq is not fit to function independently merely because the machinery of government may not run quite so smoothly or so efficiently as in some advanced or more highly developed state.

To clinch the matter, the High Commissioner offered, with the glib assurance of a salesman hawking an untried gimmick, an impressively watertight guarantee which soon afterwards turned out to be as leaky as a sieve:

His Majesty's Government fully realized its responsibility in recommending that Iraq should be admitted to the League . . . Should Iraq prove unworthy of the confidence which had been placed in her, the moral responsibility must rest with His Majesty's Government.[8]

Soon after having attained full, unfettered sovereignty, Iraq lurched into disorder and violence. This remained its usual condition until the *coup d'état* of 14 July 1958 which, as the term goes, liquidated the royal family and the fourteen-times Prime Minister, Nuri al-Saʿid. The violence had two aspects. In the first place, the Iraqi government ruled over an extremely heterogeneous population which had no reason to feel loyal to a regime run by a collection of hitherto largely unknown figures, at whose head stood a foreigner who had no roots in the country. The other side of this coin is that the rulers, unsure of their authority and fearful of any challenge to it, dealt brutally with any sign of disaffection. The Kurds, led by a tribal leader, Mulla Mustafa Barzani, rose in 1935. Barzani rose again in 1943; on British advice the Government tried to negotiate with him, but the negotiations came to nothing, and an army expedition was sent in 1945 which brought the uprising to an end for the time being. Barzani fled to Iran. Again in 1935 and 1936 in protest against army conscription decreed by Baghdad, southern tribes rose in rebellion. Conscription had been, from the 1920s, a great ambition of the regime. During the mandate, this ambition was resisted by the British and was also vocally opposed by the Shiite tribal leaders. Shortly after independence, when British objections could be disregarded, a conscription law was enacted. Dislike of such a law on the part of southern tribal leaders had by no means disappeared, hence the uprisings which were suppressed by the new conscript army. The Yazidis, a small esoteric sect in the north-west, also objected to conscription on religious grounds and pointed out that they had been exempt from it under the Ottomans. They were also very harshly dealt with: seven headmen of villages and two Christian notables from Mosul were tried by court-martial and hanged.

Perhaps the most frightful of the incidents which have punctuated relations between the rulers of the Iraqi kingdom and their subjects is the one which inaugurated the annals of the new state, the incident, namely, of the 'Assyrian Massacre'. The Assyrians, as they came to be called by Western missionaries in the nineteenth century, were Christian tribesmen long established in the mountains of Southern Anatolia. During the First World War, at Russian instigation, they rose against their Ottoman masters. Under very difficult conditions they had to fight their way out of their native lands, and at last reached British-occupied territory in Mesopotamia. These refugees supplied their new protectors with levies which proved courageous and efficient, and they were established in settlements in the environs of Mosul. Like the Kurds, they were taken aback by, and made fearful of, the prospect of living under an independent Baghdad government. They made representations to the British Government before the end of the Mandate, but were forcefully told that they could not be granted the communal autonomy they desired. In response to a petition sent on the Assyrians' behalf to the League of Nations, the British Government declared:

They are satisfied that, upon the establishment of Iraq as an independent state, member of the League of Nations, there will be no need for any special discrimination in favour of racial and religious minorities beyond such general guarantees as have been taken in the past from other candidates for admission to membership of the League.[9]

Shortly after the grant of independence, however, the Iraqi army was involved in a fracas with a band of Assyrians on the frontier with Syria. In order to teach them a lesson they would not easily forget, the army fell upon an Assyrian village and massacred the women and old men who were its inmates. The army was received with enthusiastic plaudits on its triumphant return to Baghdad.

As has been said, there was another aspect to the violence which shook Iraq after independence. In quest of power and office, the Baghdad politicians brought into play southern tribal leaders who were tempted with offers of land and privileges if they rose in rebellion against the existing government. The cycle of disorders and rebellions in the south began shortly after Faysal's death in September 1933. Discontent among southern notables and tribal leaders was not slow in being manifested. It led to the resignation in August 1934 of a cabinet which Jamil al-Midfaʻi had formed the previous February. He was succeeded by Ali Jawdat al-Ayyubi who dissolved the Chamber of Deputies and announced new elections. He packed the new Chamber with his own supporters among the tribal leaders. This naturally aroused the discontent of those whom he had disappointed. They banded together, appealed for support from the highest Shiite religious authorities, and threatened to bring down the Government. Ayyubi resigned in February 1935. His predecessor, Midfaʻi, replaced him. The tribal leaders, encouraged by Midfaʻi's rivals in Baghdad and abetted by prominent religious figures, began an armed rebellion. Midfaʻi mobilized other tribes in his support, but these proved weaker than the rebels. Midfa'i resigned after some ten days in office. The political allies of the rebellious tribes now took office in a cabinet headed by Yasin al-Hashimi. His rivals, however, instigated rebellions against him, which erupted in May 1935 and for a whole year intermittently involved various tribes with their different grievances. Hashimi sent the army against the rebels and declared martial law in the area of the rebellion, which was administered with increasing harshness. One of the army commanders sent to deal with the rebellion was Bakr Sidqi who had distinguished himself in the Assyrian Massacre. Here too he followed, it would seem, the same method, his soldiers being ordered to burn the produce on rebels' land, to destroy their houses, and indiscriminately kill those wounded or captured in battle, as well as non-combatants. The south was at length cowed and, as they say, pacified.

Bakr Sidqi, however, did not see why he should do Hashimi's bidding when he himself, commanding a whole division which had sowed terror both in the north and in the south, had the power to become the master. He plotted with another divisional commander and with one of Hashimi's fellow-plotters, Hikmat Suleiman, who had been disappointed when Hashimi did not include him in his administration. There ensued a military *coup d'état* in October 1936 which forced Hashimi to resign and go into exile, together with Nuri al-Sa'id, his foreign minister. Hashimi's defence minister, Ja'far al-'Askari, another ex-Ottoman officer who had gone over to the Sherifian cause during the war, was murdered at Bakr Sidqi's orders.

Hikmat Suleiman, then, succeeded Hashimi as prime minister, but real power lay with Bakr Sidqi who now became chief of staff. His hand remained heavy over the population: a rebellion against conscription in the middle Euphrates in 1937 was put down with the customary undiscriminating ferocity. Bakr's reign, however, did not last long. In August 1937 a plot by army officers led to his murder at the hands of a corporal who was put up to do the deed. Hikmat fell and Midfa'i succeeded him.

The cycle of tribal rebellions was followed by the cycle of military *coups d'état* which Bakr Sidqi inaugurated. The military conspirators who had encompassed his murder were dissatisfied with Midfa'i, and began to plot against him. They were instigated, aided, and abetted by Nuri al-Sa'id who had returned from exile following Bakr's downfall, and who was discontented at being kept out of office. The officers made ready for a military takeover and sent one of their number to Midfa'i to demand his resignation, while the chief of staff visited King Ghazi who had succeeded his father in 1933, and told him that the army demanded Midfa'i's resignation. Midfa'i fell and Nuri sat in his place.

In April 1939, Ghazi, dissipated and reckless, killed himself by driving into a lamp-post probably while drunk. His son was still a child and a regent had to be appointed. Nuri and his army supporters favoured Abd al-Ilah, Ghazi's cousin and the child's maternal uncle, against candidates favoured by Nuri's rivals. It was made known to the Parliament, meeting to designate the regent, that a mechanized force under the command of a military confederate of Nuri's was waiting in the wings. The Parliament duly designated Abd al-Ilah.

The Mufti of Jerusalem, Amin al-Husayni, had fled from Palestine in 1937, to escape arrest by the British authorities who believed that he was the leader of the Arab rebellion which began in 1936. He took refuge in Beirut where the French authorities left him undisturbed. On the outbreak of war, he feared that his freedom would be restricted since he was known to have connections with the Axis Powers and the British would now be in a better position to ask their French allies to intern him or even, perhaps, to return him to Palestine. In October 1939 he therefore moved to Iraq where Nuri's government made him welcome. He soon established close relations with politicians and army officers and became a power in Iraqi affairs, hoping to use his new influence in Baghdad to advance the cause of Arab unity and thus, indirectly, the cause of the Palestine Arabs in their struggle against the British and the Zionists. He seems to have persuaded the chief of staff, a major-general, and a colonel—all members of the military cabal which had supported Nuri since 1937—to work for Nuri's downfall and his replacement by Rashid Āli al-Gaylani, judged to be less friendly to the British than Nuri and more willing to adopt a belligerent stance toward them. These three officers visited the Regent and demanded Nuri's dismissal. Nuri, however, was too adroit for them and, supported by the other members of the military cabal, persuaded Abd al-Ilah to issue a decree stripping the conspirators of their positions and sending them into retirement.

The Mufti, however, succeeded in detaching the remaining members of the cabal from Nuri, and inducing them to transfer their support to Gaylani. The change in the military situation in Europe in the spring and summer of 1940 persuaded them that the Axis Powers would sweep all before them, and the Mufti argued that Germany would support Arab independence, do away with the Jewish national home in Palestine, and provide a generous supply of weapons for the Iraqi army. The plotters, four colonels, decided to force the Regent to keep in power Rashid Āli, Nuri's successor as prime minister. He succeeded in evading their demand by fleeing to a provincial centre where he was out of their hands. On 31 January 1941 Rashid Āli therefore felt compelled to resign. But the colonels were determined men. Two months later they extorted a resignation from the new prime minister. The Regent fled and the colonels established a military government headed by Rashid Āli. The Parliament, bowing to the *fait accompli*, deposed the Regent and replaced him by an obscure member of his family.

In this episode, the officers' ambitions became much more grandiose. They were now the embattled champion of the Arab world against British imperialism which was treacherously supporting the Zionists. Succumbing to the Mufti's seductions, they saw themselves as the single-handed orchestrators of a regional policy which would take on the British Empire and destroy its position in the Middle East. By the terms of the Anglo-Iraqi treaty of 1930 the British had the right in time of war to land troops in Iraqi territory. They decided to exercise these rights, but Rashid Āli and his officers determined to whittle them down. It soon came to a clash with British forces sent from India which had landed in Basra, and with a British air-force base to the west of Baghdad. Hostilities broke out on 2 May 1941. British forces were reinforced from Palestine and Transjordan. The Germans, however, provided no effective help, and by the end of the month the Iraqi forces were defeated. Rashid Ali, the Mufti, the four colonels, and those Iraqi politicians who had thrown in their lot with them, fled to Iran. The colonels were in due course captured and returned to Iraq where they were condemned to death and executed. The last one to suffer this penalty was Colonel Salah as-Din al-Sabbagh, who was hanged in 1944 at the main gate of the Ministry of Defence in Baghdad. The Regent's mutilated corpse was strung up on the same spot by the mob following the military *coup d'état* of July 1958.

The failure of Rashid Āli and his military supporters put a stop for some seventeen years to military intervention in politics. Between 1941 and the end of the war, the country was under British military occupation, and during this period the Regent and those politicians who had supported him against Rashid now felt free to purge the army in order to ensure, so they hoped, that officers would no longer intervene in politics. The constitution was also amended to give the king the right to dismiss governments. Power then became even more concentrated in the hands of the royal court and the politicians who attracted royal favour. Of these the most prominent was now Nuri al-Saʿid who, at the time of Rashid Āli's *coup d'état*, had thrown in his lot with the Regent and the British, and who thereafter had no more need of the intrigues and conspiracies to which he, like his fellow-politicians, had resorted in the unbridled pursuit of power.

After the explosion of 1941 Iraqi politics remained the appanage of the same narrow groups of Baghdad politicians who followed one another in ministerial positions in haphazard and short-lived combinations. However, even though military intervention in politics was terminated—for a time at any rate—other elements of turbulence began to be discerned after 1945. The ideological effervescence in schools and colleges which had been increasingly evident in the 1930s continued to be manifest. Arab nationalism was, as before 1941, zealously disseminated but its very dissemination paradoxically weakened a regime ever more widely believed to be a creature of the British, who had set it up in 1920 and saved it from destruction in 1941—and British imperialism was, by definition, the determined enemy of Arabism. During the war, also, leftist and communist ideologies began to be more familiar, and further to radicalize the young, and incline them to political activism and popular agitation. Thus in 1946 a Communist-inspired demonstration against the British Embassy in Baghdad clashed with the police and one demonstrator was killed. Shortly afterwards strikers against the British oil company in Kirkuk were in a confrontation with the police who opened fire on them as they fled, killing sixteen men, women, and children.

Much more serious were the disturbances of January 1948. In that month the Prime Minister, Salih Jabr, after consultations with the leading political figures, and with the Regent's agreement, signed at Portsmouth a new Anglo-Iraqi treaty to replace the 1930 one due shortly to expire. News of the signature led to a wave of demonstrations instigated and orchestrated by communists and leftists and by nationalist figures who had never reconciled themselves to the defeat of the good old cause in 1941. The strikes, which went on intermittently throughout January, led to violent clashes between the students, the mob, and the police in which twenty were killed and seventy-seven wounded. The Regent lost his nerve, disowned the treaty and dismissed Salih Jabr. The monarchical regime never recovered from this blow to its prestige.

The regime sustained another powerful blow of the same kind in 1952. The troubles began at the end of October 1952 with a dispute over examination regulations between the administration and the students of the college of pharmacy, which led to their striking. Though the authorities gave in, there was a fracas in the college which led to a renewed strike. This was joined by students from other colleges. The government ordered all schools and colleges closed, but the student demonstrations went on and were soon joined by a large mob. A police station in Baghdad was besieged and a policeman lynched. The demand was now for direct elections to be substituted for the indirect ones hitherto in force. There were cries of 'Down with the monarchy' and 'Long live the republic', made the more ominous by the recent deposition of King Faruq of Egypt at the hands of a military junta. The army was called in, martial law was declared and troops fired on the rioters, killing over twenty. The Regent dismissed the administration, headed by a political nonentity, Mustafa al-'Umari, and appointed the chief of staff as prime minister. The new Prime Minister conceded direct elections. The change did not make the deputies more representative or Parliament more significant than hitherto.

The monarchy, its institutions, and the political figures who derived from them power and profit were to last less than six more years. In this period, which saw the heady influence of Colonel Nasser reach its apogee in the Middle East, the regime became increasingly delegitimized in the eyes of the educated classes and of the military officers. There was a surface glitter imparted by a great increase in oil royalties which the regime obtained in 1952. But disaffection, albeit hidden or suppressed, continued to increase—disaffection towards Abd al-Ilah who, since his nephew Faysal II attained his majority, had become the heir to the throne, and who was very much the power behind the throne; towards a regime which was seen as oppressive, corrupt, and an instrument of imperialist designs; and towards a political class which was believed to set its face against all that was alive and vibrant in Arab politics. Thus, in spite of the purges which had been carried out after 1941, and of an ever-vigilant surveillance by a swarm of informers inside the armed forces, a network of officers was established which engaged in successive conspiracies against the regime. On 14 July 1958, one of the conspiracies, led by Brigadier Abd al-Karim Qasim, finally succeeded. In the course of one morning the whole political structure jerry-built in 1921 crashed to the ground, and its main pillars murdered or imprisoned. The constitution and the parliament, such as they were, were abolished and Iraq became a republic governed at his whim by a military officer.

3. Syrian Politics 1920–49

The country known today as Syria became, like Iraq, a mandated territory following the First World War. France was designated as the Mandatory Power. France had had traditional interests in this portion of Ottoman territory and, in the course of the war, the Allies agreed among themselves that this area, together with what is now known as the Lebanon, should be one where French influence and French desiderata would be paramount. In October 1918 the British, having borne the brunt of the fighting against the Ottoman armies, occupied the whole of the area which France claimed according to the Asia Minor Agreement (commonly known as the Sykes–Picot Agreement, from the names of the negotiators) signed in May 1916. The British Government, however, had had second thoughts about this arrangement and attempted to do away with French claims. To this end, Allenby, the British Commander-in-Chief, forbade all his troops, with the exception of Faysal's Northern Arab Army, to enter Damascus, and proceeded to install Faysal as the head, in effect, of an Arab administration. The obvious calculation was that this administration would claim to govern in the name of national self-determination, thus doing away with French control over the territory—control which would make France a powerful rival to the British empire in the Middle East.

The gambit failed when, in September 1919, the British gave up the idea of supporting Faysal against the French, and agreed to withdraw their troops from all the area of the Levant which the Sykes–Picot Agreement had allotted to the French. In April 1920, the Allied Supreme Council awarded France a Mandate for Syria and Lebanon—an award subsequently ratified by the League of Nations. A large contingent of French troops had meanwhile been sent to Beirut in replacement of the British forces which had evacuated the Lebanon. Faysal had been proclaimed King of Syria by his followers the previous March, and was increasingly pressed by them to offer resistance to the French who themselves were bent on imposing their control over the new kingdom. Faysal, weak and hesitant, was unable to resist the pressures, or to conclude with the French an agreement which they considered acceptable. It soon came to a clash. On 24 July French troops clashed with, and defeated, Faysal's forces, entered Damascus, and put an end to the short-lived kingdom. A few days later Faysal, having given up hope of an accommodation, and in response to French threats, left Syrian territory for Haifa where a memorial column, still standing, commemorates the official welcome accorded to him in person by Sir Herbert Samuel, the British High Commissioner for Palestine.

The territory which the French now had to govern was by no means settled and peaceable. The short-lived Sherifian regime had been disorderly and disorganized. Its government, Miss Gertrude Bell reported after a visit to Syria in October 1919, was perceptibly worse than that of the Turks, with worse public security, more venality in the courts, and the usual corruption among high officials. Faysal's followers in Damascus and the provinces had fomented disorders in Mesopotamia and against French garrisons in the Levant, and against areas whose populations were suspected of sympathies with the French. The frontier with what was to become Turkey was by no means defined, and it was in the interest of the Kemalists in Ankara to create trouble for the French. The Sykes–Picot Agreement had awarded Cilicia to France. The Kemalists, however, were determined that this territory should remain under Turkish sovereignty, and were thus bent on creating difficulties for the French in the Levant, either by direct action, or co-operation with Faysal. The Druse, again, would by no means be easy to control. At the best of times, the Ottomans were able to police the Druse Mountain only imperfectly. The war and the Sherifian regime created conditions such that the Druse chieftains were more unwilling than ever to bow under the yoke of a central government.

The Syria of the mandate was made up of the largest part of the Ottoman province of Aleppo, of parts of the province of Damascus, and of the northernmost bit of the province of Beirut. Though unlike Iraq in being predominantly Sunni Arab, the Syrian population was nonetheless quite heterogeneous. In the south there were the Druse, who practised an esoteric religion originally stemming from Shiism and who predominated in the Mountain of the Druse and in the Hawran; in the north-west lived the Alawis, followers of another esoteric cult composed of eclectic elements derived from Shiism and gnosticism, and denounced by Sunni heresiographers as unbelievers and idolaters; to the north-west of them lay Alexandretta and its hinterland, inhabited by Turks, Armenians, Alawis, and Sunni Arabs. The Jazira, in the north-east contained Kurds, Christians, Arab nomads, and Assyrian refugees whom the French authorities allowed to resettle there following the Assyrian Massacre of 1933.

The policy of the French Mandatory, in great contrast to that of the British in Iraq, was to take account of this heterogeneity when setting up constitutional and administrative arrangements. In so doing, the French were following the injunction of Article 1 of the Mandate which stated that the 'Mandatory shall, as far as circumstances permit, encourage local autonomy'. This requirement chimed in, of course, with French interests. In occupying Syria they had clashed with Faysal and his predominantly Sunni Arab followers who, the French feared, would always be antagonistic to the mandatory regime. Furthermore, the French considered themselves to be the protectors of the Maronites in the Lebanon, and of the Catholics generally in the Levant. They would thus wish to shield them from the hegemony which Muslims might be tempted to exercise, both by reason of their superior numbers and by the habit of domination over non-Muslims which had become ingrained during the long centuries of Muslim rule. French policy also tried to take into account, and make provision for, the great differences and rivalries which existed between the two main Syrian centres, Damascus and Aleppo. Damascus had always looked eastward to the Syrian desert, and southward to Arabia, while Aleppo was a great mercantile entrepôt which looked northward and eastward to Anatolia and Kurdistan, and westward to Alexandretta, Latakié, and Mediterranean networks of trade.

Thus, in the two years following the occupation, the Mandatory established separate states of Damascus, Aleppo, and of the Druse country and a separately governed territory for the Alawis. In 1924, the arrangement was modified by amalgamating Damascus and Aleppo in a unitary State of Syria. The instrument of the Mandate approved by the League of Nations in July 1922 required the Mandatory to frame, within a period of three years from its coming into force, an organic law for the mandated territory. This, however, proved impossible owing to the outbreak of disorders in Druse territory which spread to other parts of Syria, and which went on intermittently in one area or another.

The initial outbreak in the Druse mountains was led by a chieftain, Sultan al-Atrash, who was much discontented with the brusque methods and the heavy hand of French administrators, particularly following the arrival at the beginning of 1925 of a new High Commissioner, General Sarrail. Druse chieftains were alienated by the fear that their power over their followers was being eroded and was passing to officials of the Mandate and to the Representative Council which the French had set up in Druse territory in common with other Syrian states. The kind of government which works according to bureaucratic routine; which levies and enforces the collection of taxes; which engages in extensive public works for which it demands compulsory labour, was quite unfamiliar. There was complaint and disgruntlement in a population which was anyway warlike, accustomed to obey its traditional leaders, and difficult to govern by a central authority whether Ottoman or French. Atrash's uprising was sparked off by Sarrail summoning five Druse chieftains to Damascus on the pretext that the authorities there would consider Druse demands, and having them arrested and sent to exile in Palmyra in the Syrian Desert.

The Druse rebellion began on 18 July. On 2 August, a column of French troops proceeding to relieve the besieged garrison in Suwayda, the capital of the Druse Mountain, was ambushed. Eight hundred men were lost and the Druse fighters captured a large quantity of arms and ammunition. This victory inflicted a great blow on the prestige of the Mandatory and greatly increased Atrash's self-confidence: the more so that

French troops were thin on the ground and that France was also facing a rebellion in the Rif mountains in Morocco. The setback suffered by the French troops near Suwayda had immediate repercussions which affected the whole of Syria. Political figures in Damascus, Sunni members of the intellectual and official classes and far removed from Sultan al-Atrash's outlook and loyalties, saw in the Druse uprising an opportunity to shake the French mandate to which they had never been reconciled—and the less so that they looked back with nostalgia to Faysal's short-lived regime and the vista of political independence which they thought it had opened up, but which the French had so ruthlessly closed off. One of these leaders, Abd al-Rahman al-Shahbandar, met Atrash and agreed with him on the proclamation of a Syrian National Government to be headed by Atrash himself. More to the point, Druse bands infiltrated the Ghuta, an area surrounding Damascus, thickly planted with trees and bushes and intersected by a network of irrigation canals. Together with sympathizers from the Ghuta villages and marauding bands out for loot, they penetrated briefly into parts of Damascus in August and October 1925 and in April 1926, and on the last two occasions the French bombarded quarters which they had invaded. Again, in Hama, an uprising was led by an ex-Ottoman officer who deserted from the Syrian Legion, an auxiliary formation set up by the French. Hama slipped for two days out of the control of the lawful government. Disorders also occurred in Aleppo and its environs.

By the first months of 1927, however, the disorders had gradually died down, and the Mandatory set about making arrangements for the framing of an organic law as required by the Mandate. In April 1928 a Constituent Assembly was elected which met the following June in order to begin drafting a Constitution. In this Assembly, there was a clear majority of what might be called unpolitical representatives whose instinct was to go along with the wishes of the administration. In cities, however, nationalist notables were able to muster support in various quarters where they had local influence, and won many seats. In the Assembly they were the only organized group, and they succeeded in electing as President Hashim al-Atasi, of the Homs landowning family, who had served as Prime Minister in Faysal's regime. The High Commissioner left the drafting of the Constitution to the Assembly. The draft proposed that Syria should be a republic with a single-chamber parliament, a president, and a council of ministers. This was the usual constitution with which League of Nations-mandated territories and ex-colonies of Western empires were endowed then and later—and which, as the sequel was to show, had little or no significance. Under the influence of the nationalist members, the Assembly included in the draft articles which went against the terms of the Mandate: including in 'Syria', Lebanon, which was a separate territory under French mandate, as well as Palestine and Transjordan which were under a British mandate; also clauses giving the Syrian government powers to conduct foreign relations, organize the armed forces, and declare martial law. The provisions were obviously unacceptable to the Mandatory. The Assembly would not agree to delete or qualify them. Eventually the High Commissioner suspended and then dissolved the Constituent Assembly and promulgated by decree the constitution as drafted, less the unacceptable articles, in May 1930.

Elections under the new Constitution took place in January 1932. In spite of disorders and intimidation fomented by the nationalists now organized in a party, the National Bloc, the elections resulted in a considerable non-nationalist majority. By then, however, Iraq, which had also been governed under a mandate, was on the point of attaining full sovereignty as well as membership of the League. This example, and the very logic of the manadatory idea, namely that countries under mandate were to be prepared by the mandatory for self-determination, meant that the French would be under constant pressure to do in Syria what the British had done in Iraq. Towards the end of 1932, the French began negotiating the terms of a treaty with the Syrian government which had been formed following the elections. The draft treaty provided for a Franco-Syrian alliance, for the progressive transfer of powers from the Mandatory to the Syrian Republic over a period of four years and, at the end of this period, for Syrian admission to the League as a sovereign state. The draft treaty, however, also provided that the Druse and the Alawite States should, for the time being at any rate, remain separate from the Syrian Republic. With the agreement of his cabinet, the Prime Minister signed it. Demonstrations and rioting against the treaty were instigated by the Bloc in Damascus and other cities which, in the judgement of the High Commissioner, made impossible its proper consideration by the Chamber of Deputies, and he withdrew it from their cognizance. The Chamber was suspended, and with nationalist agitation and rioting going on, in November 1934 the suspension became *sine die*.

In November 1935 a nationalist leader from Aleppo, Ibrahim Hananu, died. The ceremonies commemorating him on the fortieth day of his death served as occasion for the National Bloc to renew the agitation against the Mandatory, and in favour of an independent and unitary Syria in which the Druse, the Alawis, the territory of the Jazira and Alexandretta—the latter two areas having mixed populations in which Kurds and Christians in one case, and Turks in the other, predominated— would all be governed from Damascus by what necessarily would be an Arab Sunni government. Demonstrations, strikes, and riots orchestrated by the Bloc during January and February 1936 followed. The High Commissioner at first refused to give in to the agitation, but at the end of February, presumably on instructions from a new, left-of-centre government in Paris, he met the Bloc leaders. It was agreed that negotiations for a treaty would be resumed, and that a Syrian delegation would proceed to Paris for this purpose. The delegation was composed mostly of Bloc leaders. In agreeing to negotiate with them, the French were bowing to agitation and violence, and also deferring wholly to the claim of the National Bloc that they were the only genuine representatives of the people of Syria. There was, of course, no evidence that this was the case. Like the British taking Zaghlul and his Wafd as their own valuation, the French too now similarly accepted the Bloc at their own valuation, and in so doing conferred on them a representativeness which the sequel showed to have no reality or meaning.

A month after the Bloc delegation arrived in Paris, the socialist Popular Front formed a government in France, remaining in office until June 1937. This government overturned the assumptions on which the policy of the Mandatory had been based since 1920. It accepted the Bloc's demand that an independent Syria should be a centralized state governed from Damascus. The Mandate, it was also agreed, would terminate three years after the coming into force of the treaty. The Bloc delegation returned in triumph from Paris. Elections were held in November 1936. The Bloc naturally enjoyed a sweeping victory, and took the reins of government. However, the Popular Front Government in Paris was succeeded by governments which did not feel disposed to follow the same policy, and the Foreign Affairs Committees both of the Chamber of Deputies and the Senate were opposed to ratification of the 1936 Treaty which thus fell to the ground.

Meanwhile, Syria under National Bloc rule was a country of unease and disturbance. Neither Alawis nor Druse nor the various groups in the Jazira felt confident that their interests would be respected by the new regime which had proceeded to monopolize high administrative positions both in Damascus and the provinces to the benefit of its supporters. Corruption, nepotism, and maladministration became rife. Druse, Alawis, Kurds, and Christians in the Jazira loudly complained about the new regime, and its heavy-handedness. The complaints occasionally led to armed confrontations between local groups, and the new rulers and their supporters. The Bloc also established a youth movement which formed paramilitary squads, the Steel Shirts, in imitation of contemporary Fascist and Nazi practice. The Steel Shirts engaged in brawls with rivals and opponents of the Bloc, whether nationalists who accused the Bloc of weakness in its dealings with the Mandatory; or youth movements set up in defence of the Druse, the Alawites, and the Jazira Christians. The National Bloc was also opposed for its lack of nationalist zeal by Shahbandar who had now come back from the exile to which he had been constrained by his part in the Druse disorders of 1925–6. Following its triumph in Paris and in the elections of November 1936, the Bloc, in reality an unstable coalition of notables, showed itself to be riven by personal jealousies, and by rivalries, the most serious and deep-seated being the traditional one between Damascus and Aleppo.

During the first half of 1939, the Bloc Government was visibly in disarray. Public order deteriorated, with demonstrations, parades, strikes, and riots. In July the High Commissioner re-established the status which the Druse and Alawi territories and the Jazira had enjoyed before the abortive treaty. The Bloc leader, Hashim al-Atasi, had been elected President of the Syrian Republic following the electoral victory of November 1936. Atasi now resigned. The High Commissioner suspended the Constitution and dissolved the Chamber, putting the Government in charge of a council of directors-general of the various ministries.

The outbreak of world war in September put a stop to further negotiation over the Syrian Mandate. In June 1940, the Third French Republic was destroyed through military defeat at the hands of the Germans, and was replaced by the Vichy regime. The Mandatory authorities gave their allegiance to the new rulers. These, in order to obtain concessions from the occupier, allowed German planes to use Syrian airfields and gave permission for military supplies to be sent to Rashid Āli in Iraq, during May 1941. These actions disquieted the British who were afraid that Axis forces might become established in the Levant and menace, from yet another direction, the British position in the Middle East to which the *Afrika Korps* under Rommel in Tripolitania was already a formidable threat. The Free French were also pressing for an invasion of the Levant in the hope that the bulk of the French forces stationed there would go over to de Gaulle. In June 1941 the Allied invasion took place and, in due course, the forces loyal to Vichy surrendered. Very few of them, however, opted for the Free French cause, the vast majority choosing to go back to France. De Gaulle claimed that his movement was the legitimate government of France and had, as such, the right to administer the Mandate. Given that his forces were much inferior to those deployed by the British, he perforce became, and was seen to be, Britain's junior partner. Furthermore, when the invasion began, the Free French published a proclamation in which it was declared that they were coming 'to put an end to the Mandate' and proclaim the mandated territories 'free and independent': 'Your independent and sovereign status will be guaranteed by a treaty in which our mutual relations will be defined'.[10] These statements might be taken to show that the Free French were simply continuing the policy which had eventuated in the abortive treaties of 1932 and 1936. There was, however, a difference. Before the downfall of the Third Republic France had been a free agent. This was now no longer the case. The Free French proclamation was accompanied by a British one which endorsed, and underwrote, the French promises and in effect guaranteed that the promises would be kept. It was the judgement of the British, in Churchill's words, that everything should be done to meet 'Arab aspirations and susceptibilities', and that 'Arab independence' was 'a first essential' with which 'nothing must conflict'.[11]

This was the policy followed by the British representative in the Levant, Sir Edward Spears. As early as the Spring of 1942 the British began to press for elections to be held in Syria and the Lebanon on the score that Egypt and Iraq were demanding elections in these two countries, because elections formed 'the indispensable criterion' of true independence. Rommel's lightning advance in the summer of 1942 through Cyrenaica to the gates of Alexandria muted these demands for the moment, but the pressure was renewed when the victory at El Alamein removed the Axis threat. The Free French reinstated the Constitution of 1930, and elections were held in July 1943.

In the circumstances, the victory of the Bloc was a foregone conclusion. So was the eviction of the French from the Levant, which took place two years later. In May 1945, the Mandate being still formally in force, the French attempted to assert their authority over various points of dispute with the Syrian government. Strikes, demonstrations, riots in various Syrian cities, including Damascus, led the French to use military force in order to quell the disturbances. French forces, however, were not alone in Syria. The British troops were also stationed there. Acting on instructions, their commander demanded and obtained that all French troops should be confined to their barracks. The Syrian Mandate was thus made utterly meaningless.

During the four years which followed, the Bloc governed Syria—in a manner of speaking. As in Iraq during the monarchy, ministries followed upon one another hapazardly, at the whim of the very small group who played the political game. The Bloc, which had purported to speak with the unanimous voice of the Syrian people, now split into two factions: the People's Party based in Aleppo and the National Party based in Damascus. This division might have introduced some checks and balances in the government were it not that the executive in this centralized state was very powerful, not to say omnipotent, and that official patronage was by far the most important element in deciding the outcome of elections. The character of the parliament in independent Syria is well brought out in the testimony of two Syrians in public life:

the sober vocabulary of politics—terms such as 'deputy' and 'opposition', 'President of the Chamber', and 'Foreign Affairs Commission'—failed to cloak the lamentable confusion reigning in the Syrian parliament. 'I look around me', writes Habib Kahaleh in his *Memoirs of a Deputy*, 'and see only a bundle of contradictions' . . . Men whom nothing united, sharing no principles, bound by no party organization, elected to parliament by some mysterious travesty of free and unhampered elections; some were illiterate, others distinguished men of letters; some spoke only Kurdish or Armenian, others only Turkish; some wore a *tarbush* [fez], others a *kafiyeh* [the bedouin head-dress]; townsman and bedouin. It was all play-acting.[12]

Khalid al-Azm (1903–65) descended from an aristocratic Damascene family, man of affairs, diplomat, minister, and many times prime minister, was a much more important figure than Habib Kahaleh, and much more intimately acquainted with public affairs in Syria. In his *Memoirs* he draws a comparison between consultative commissions, drawn from various groups which were appointed by the French in order to seek expert advice on public affairs, and elected assemblies. The elected assemblies, he argues, are composed of deputies chosen by the party in power which uses various well-known methods to enforce their election by the people. The elected assemblies from 1937 to 1939 and from 1943 onwards have always passed a vote of confidence in favour of a newly formed government, and governments retain this confidence until something outside the ambit of the assembly takes place, and the prime minister then resigns. In most cases the assembly is not aware of the reasons for the resignation. A consultative commission, composed as it is of experts in various subjects, is

a thousand times better than a representative assembly made up of members of political parties, or rather the existing partisan groupings. We cannot call such groups a political party, since their programmes are unclear, and since they are based on personal sympathies. The assembly is composed of representatives of localities most of whom are illiterate, who attend only to advance their own interests, or to follow up the problems which their own electors have. As for legislation, and the consideration of the economic, sanitary, and civilizational conditions of the state, experience has shown . . . how little interested the representatives had been in these public affairs. Interested criticism, attacks on those carrying on the government, the abuse of inflated oratory for the sake of cheap popularity: all this sums up the bitter experience of the country during the years of parliamentary government.[13]

In July 1947 new elections were held. The electoral law had been changed shortly before and direct elections substituted for the indirect elections which had been the rule. This change, however, did not signify that elections mirrored more accurately the wishes of the voters. As Azm wrote, direct elections were both comic and scandalous, since 'the dead gave their vote even before the living', and boxes stuffed with falsified votes were substituted for those in which the voters had deposited their voting papers. The largest group in the new assembly consisted of so-called 'Independents' whose votes were manipulated and controlled by ministers and, above all, by the President of the Republic. In 1947 the President was Shukri al-Quwatli, a Bloc leader who had been elected for a five-year term in 1943. The Constitution forbade an incumbent President to be re-elected. Quwatli, however, induced the assembly to amend the Constitution and allow his election for a further term. This was done and Quwatli was elected for a second term in April 1948.

Profiteers rejoiced and scrambled for their share of the loot; the old team was back in power for another stretch; there was much talk of trafficking in import licences and in the agencies for imported goods. Quwatli himself . . . sat on top of an edifice of nepotism and mismanagement.[14]

The edifice was to crash down not long afterwards. In May 1948 the Syrian army, in common with other Arab armies, invaded Palestine which the British had evacuated, in order to prevent the establishment of a Jewish state. Badly led and poorly equipped, Syrian forces acquitted themselves miserably. Officers complained that this poor showing was the outcome of corruption and incompetence among the civilians. It may have been in order to rebut these accusations that Quwatli sought to put the blame for the corruption on the army officers themselves. On a visit to the front-line, early in 1949, Quwatli discovered that cooking fat supplied for the troops was of inferior quality. Investigation established that the cooking fat was not made of sour milk, as was usual, but from bone waste. Quwatli ordered the arrest of the chief supply officer, a colonel who had been appointed to this post by his friend the Chief of Staff, Colonel Husni al-Za'im. Za'im, whether because he was himself implicated in the cooking-fat scandal and feared exposure, or because he took offence at civilians attempting to accuse the army of corruption, when they themselves were the guilty parties, organised a *coup d'état* in March 1949. He put in prison Quwatli and his Prime Minister, Khalid al-Azm, and himself took power. He held a referendum the following June which elected him President of the Republic. The following August, he was toppled by another Colonel, Sami al-Hinnawi, who arrested Za'im at dawn, had him tried by a court-martial held at a roadside and immediately executed, together with his Prime Minister, Muhsin al-Barazi, who had been Quwatli's personal assistant and speech writer. Hinnawi became President of a Revolutionary Council, but in December he was overthrown by another Colonel, Adib al-Shishakli, and went into exile in Beirut where he was murdered the following year in revenge for Barazi's death, which itself was no better than murder.

Shishakli did allow the cabinet drawn from the People's Party which had been set up under Hinnawi's wing to continue in office for a while—and this cabinet was followed by other short-lived administrations. On 28 November 1951, following disagreements between the People's Party and Shishakli, a member of the People's Party, Maʿruf al-Dawalibi, succeeded in forming a cabinet. Shishakli did not find it acceptable and that same night ordered the arrest of Dawalibi, his cabinet, and other prominent members of the People's Party, including Nazim al-Qudsi, its secretary-general and a former Prime Minister. Shiskhali then conferred the triple dignity of Head of State, Prime Minister, and Minister of Defence on Colonel Fawzi Salu, who became a Field Marshal. However, in June 1953, Shishakli made himself Prime Minister and a popular referendum elected him President. He ruled until February 1954 when a Colonel, a Major, and a Captain carried out a *coup d'état* which toppled him. He went to the Lebanon, then to Saudi Arabia and France, before emigrating to Brazil where a Druse assassinated him in revenge for having bombed Druse villages during a Druse uprising shortly before his overthrow.

These events show that 1949, the year of the three colonels, marked the virtual end of constitutionalism in Syria. Even though there were civilian cabinets in office between 1954 and 1958, there was no doubt that the substance of power lay with army officers. It was they, as well, who forced a union with Egypt under Nasser in 1958, and they again who undid it in 1961. It was they also who, from conspiracy to conspiracy between 1961 and 1964, finally made sure that Syria was for the soldiers to govern. Khalid al-Azm's memoirs contain a notable passage which sums up the political experience of Syria between the elections of 1943 and the union with Egypt in 1958, but what Azm writes applies equally to the years which followed.

I consider myself one of those who were responsible, wholly or partially, for what took place in Syria between 1943 and 1958 . . . But the prime responsibility belongs to the army, its general staff and its officers, commissioned and non-commissioned. This is because it was the army which had control over the fate of the country since the *coup d'état* of 30 March 1949, when a class of young men in consequence gained control of power. They had not been able to gain ordinary school certificates, and thus applied to the Military School at Homs. The School required them to attend for only two years, and made it easy for them to acquire its diploma. An officer graduating thence with a single pip on his shoulders would begin to think that it was in reality a star scintillating like a jewel. He would walk the earth with conceit, look upon society with enmity and envy, and upon civilians with contempt. In his eyes they were all traitors, mercenary agents, and feudalists. What influenced the formation of their character was that they had been imbued with a spirit of hatred, jealousy, and rebellion against the prevailing social life, especially that their origins lay with groups which circumstances did not favour with any ease of life.[15]

What Azm said of the Syrian officer corps applied *mutatis mutandis* to Egypt and Iraq where constitutionalism also led to, and ended with, military takeovers.

4. Politics in the Lebanese Republic 1920–1975

Along with Syria, the French administered another Mandate, that of the territory to the west of Syria known as the Lebanon. This territory was, in a great many respects, very different from its eastern neighbour. The central area of what is today the Lebanese Republic was known in the nineteenth century as Mount Lebanon. It was indeed a mountainous area inhabited mainly by Druse and Maronite Christians who were affiliated to Rome. In the nineteenth century and even earlier, the Maronites were numerically predominant over the Druse. According to an estimate dating from 1861 there were 25,000 Druse against 225,000 Maronites. The estimate by and large finds confirmation in a census of 1913 which shows that Mount Lebanon contained some 242,000 Maronites and some 47,000 Druse. Taking the population as a whole, the various Christian communities of Mount Lebanon formed three-quarters of the population, and the Maronites constituted two-thirds of the Christians.

In spite of numerical inferiority, Druse chieftains ruled the Mountain for a very long time. The Druse Ma'nid emirate exercised power from 1516 to 1697. They were succeeded by the Shihabis, who were not Druse, but Sunnis, and who were the rulers from 1697 to 1841. During Shihabi rule the Maronite population expanded, and the power and prominence of the Druse diminished greatly. The Shihabis themselves converted to Maronite Catholicism, and in the twentieth century Maronites came to identify themselves very much with the Shihabi emirate as an expression of Maronite autonomy, and as establishing the traditional territorial extent of a Maronite Lebanon.

It would not be misleading to describe the Ma'nids and the Shihabis as feudal rulers who owed allegiance to the Ottoman Sultan and remitted taxes to him, but who were otherwise autonomous, and whose rule was itself based on a network of feudal landowners who gave allegiance to the emirate.

The heyday of the Shihabi emirate, which however ended with its downfall, was the long reign of Bashir II (1788–1840). He extended his authority beyond the traditional bounds of Mount Lebanon, and was therefore in his day an important player on the Middle Eastern stage. In a later generation, the era of Bashir II represented to Maronites a golden age which they were to try to re-create with the help of France. However, during the last decade of his reign Bashir became involved in international rivalries and quarrels which led to his ruin and to the destruction of the emirate regime. In 1831 the Levant was invaded by Muhammad Ali, the Pasha of Egypt, and Bashir co-operated with the Egyptians in disarming the Druse and in conscripting both Maronites and Druse into the Pasha's forces. Eventually, Muhammad Ali threatened the very existence of the Ottoman Empire when he defeated the Ottoman army at the battle of Nezib in 1839. In so doing he brought down upon himself the displeasure of the Powers, chiefly Great Britain, who were afraid that if he succeeded to the empire of the Ottomans and sat in their place at Istanbul, the (unwelcome) beneficiaries of this development would be the French who would then be in a paramount position in the area. Muhammad Ali was forced by the Powers to evacuate the Levant. Bashir's position, as Muhammad Ali's client, became hopeless. He had alienated both Druse and Maronites who united in a call to rise against him, pledging themselves 'to fight to restore their independence or die'. In October 1840 Bashir surrendered to the British and was taken into exile in Malta.

The two decades which followed the collapse of the emirate saw the Mountain plunged in troubles and disorders, which were the aftermath and consequence of the decade of Egyptian occupation, of the discord which this occupation introduced between Maronites and Druse, of intervention by various foreign Powers each with their clients, and of the attempt by a centralizing Ottoman government to impose its authority on a province hitherto more or less outside its reach. The tension and disorders involving Maronites and Druse, as well as the Maronite peasantry and their landlords, culminated in a horrific outbreak in 1860, when Druse attacked Maronites in various areas, killing and destroying. Some 10,000 were killed and some 100,000 rendered homeless. In Damascus also, Muslims massacred Christians. These events led to an outcry in Europe, particularly in France, and a French expeditionary force was sent to Beirut where it landed in August 1860. The other European Powers could not allow the Lebanese problem to be settled by the French alone. In the event, representatives of Russia, Austria, Prussia, Britain, and France formed a commission presided over by the Ottoman Foreign Minister and agreed that Mount Lebanon should henceforth be governed according to a *Règlement organique*. The working of the *Règlement* was guaranteed and supervised by the six signatories, who were joined by Italy in 1867. The *Règlement*, amended in 1864, remained in vigour until 1915, when the Ottoman government took advantage of the situation created by the world war and abrogated it.

So long as it lasted, the regime established by the *Règlement* proved very successful. Peace between the communities was preserved and Mount Lebanon prospered. The *Règlement* set up an autonomous province, the governor of which would be a non-Lebanese Ottoman Christian subject appointed by the Porte after consultation with the other signatory Powers. There was also set up an administrative council consisting of twelve members, namely two Maronites, two Druse, two Greek Orthodox, two Greek Catholics, two Sunnis, and two Shiites. The 1864 amendment increased Maronite representation to four and Druse to three, while the Greek Catholics, the Sunnis, and the Shiites were restricted each to one representative. The members of the council were to be designated by the heads of each community after consultation with the notables, and appointed by the governor. The council was to assess taxes, to administer revenues and expenditures, and to advise on questions submitted by the governor. Mount Lebanon was divided into six districts, to each of which the governor appointed an administrative agent 'selected from that religious group which predominates either in size or in the importance of its properties'. Furthermore, each district was to have a local administrative council comprising from three to six members representing the various elements of the population and the landed interests. Districts were divided into sub-districts, sub-divided into communities, each headed by a sheikh selected by the inhabitants and appointed by the governor. The *Règlement* further provided that in mixed communities each element of the population 'will have its own sheikh who will exercise authority only over his co-religionists'. The judicial organization of the autonomous province followed the same principles and took care to ensure that, at all levels, judges and justices of the peace were drawn from the various communities. The *Règlement* even took care to specify that a Protestant or a Jewish representative would sit on the higher judicial council 'whenever the interests of a member of these communities may be involved in a lawsuit'.

It is immediately apparent that the regime established by the *Règlement* draws its inspiration at once from the traditional *millet* system, and from the political experience of Mount Lebanon as it developed and became articulated under the emirate. During this long period, before the Egyptian invasion in 1831, Maronites and Druse by and large had learnt to co-exist; and given the characteristics and structure of society in the Mountain, it was neither possible for the emirs, nor to their interest, to ride roughshod over one group or another under their rule. It is in such a context that we may read the joint proclamation by Druse and Maronites during the last days of the reign of Bashir II in 1840, affirming their resolution 'to fight to restore their independence or die'. Lastly the *Règlement* reflects Western ideas of representativeness and constitutional government to be expected in a document in the composition of which European Powers played so great a part. So long as it lasted, the constitutional regime instituted by the *Règlement* was a manifest success. The vicissitudes of parliamentary and representative government before 1914 must lead us to conclude that, in the whole Middle East, only Mount Lebanon successfully practised this mode of government.

The terms of the Mandate for Lebanon required the French—as in the case of Syria—to frame, within three years of its coming into force, an organic law for the territory. In 1922 elections for a Representative Council were held. At the end of 1925, the High Commissioner transformed it into a Constituent Assembly, which proceeded to debate and approve a draft constitution in May 1926. The constitution set up a Republic to be headed by a President who would be elected by a Chamber of Deputies. In what is perhaps its most important provision, the constitution follows the spirit of the *Règlement*, and embodies the ethos of the Lebanese polity as it developed during the era of the emirate. Article 95 declares:

As a temporary measure, and in conformity with Article 1 of the Mandate [which required the Mandatory to take into account the 'rights, interests, and wishes' of the population], as well as for the sake of justice and concord, the various communities shall be equitably represented in public employment and in the composition of cabinets, provided that this causes no prejudice to the general welfare of the State.

Ten years later, when a Franco-Lebanese Treaty (which proved abortive) was negotiated, a letter from the President of the Lebanese Republic to the French High Commissioner, which formed an annexe to the Treaty, confirmed that the Lebanese Government is disposed to guarantee the equality of civil and political rights of all its citizens without exception. It is equally disposed to ensure that the different elements in the country are equitably represented in the generality of public offices. The Lebanese Government will ensure that the different regions will each receive an equitable portion of the expenditure devoted to public utilities.

Again, when, following independence, the constitution was amended in 1943 to remove from it various references to the Mandate, Article 95 was retained in exactly the same form.

The Chairman of the committee appointed to draft the constitution of 1926, Shibl Dammus, a Greek Orthodox, explained at the time the rationale of this provision. He wrote:

1. The Lebanese people is composed of a multitude of communities, each having its own religious beliefs, mentality, customs, and peculiar traditions. To reject the system of confessional representation [i.e. representation of religious communities] disrupts the equilibrium and gives preponderance to certain communities over others. Jealousy, resentment and even continuous disorder may be the outcome.

2. Representation in parliament must reflect the character of the country; and as the country is divided into several communities, it is necessary for the communities to be represented; otherwise the representation will not be faithful.

3. The country is still imbued with the spirit of confessionalism; the time has not come for these prejudices to be given up and we cannot abandon in a day a mentality which is several centuries old.

4. Confessional representation safeguards minority rights and allows no scope for recriminations.

5. Solidarity between the various communities is not yet so perfect as to enable us to disregard confessional politics.

6. The Lebanese people is not yet accustomed to give the primacy to patriotic solidarity over confessional solidarity.

7. The [religious] communities of the Lebanon are the equivalent of political parties.[16]

Confessionalism (derived from the French *confession*, meaning religious denomination), of which Dammus's Report speaks, is, in the Lebanese context, the equivalent of constitutionalism. Another member of the committee which drafted the constitution, the Greek Catholic Michel Chiha, perhaps the most acute and intelligent writer on Lebanese politics before the destruction of the Lebanese polity, repeated and amplified the same point in an article some three decades later:

Confessionalism in the Lebanon . . . is the guarantee of equitable political and social representation for confessional minorities in association . . . [it] makes above all for order and for peace . . . Lebanon is made up of confessional minorities in association. The minorities appear under a confessional label because the Lebanon has always been a refuge for the freedom of conscience. This has been possible owing to the Lebanon's geographical situation—being a mountainous land in which it has always been possible to defend oneself, and a maritime land in which it has always been easy to take to the sea . . . In spite of many mistakes and abuses, it is confessionalism which has taught Lebanon how to be tolerant . . . The Lebanese equilibrium, based on confessionalism, is not an arbitrary equilibrium. It was not created by prejudice, but rather by the need to acknowledge particularities which extend as far as those which characterize political parties. With time such differences may become attenuated and slowly disappear. At present, the Lebanon's *raison d'être* lies precisely in the confessional equilibrium which characterizes it and which shows itself primarily at the level of the legislative Power.[17]

The constitutional provision which recognized and accommodated confessionalism was given tangible reality by electoral and other arrangements. Thus, constituencies elected not a single member from among competing candidates, but rather a list of members chosen from among competing lists. The candidates figuring on each list were drawn from each of the communities inhabiting the constituency in a proportion prescribed by law. Thus the electoral law of 10 August 1950 enacted, for example, that from the constituency of Beirut there shall be elected four Sunnis, one Shiite, one Maronite, one Greek Catholic, one Greek Orthodox, one Protestant, one Armenian Catholic, two Armenian Gregorians, and one member representing all other, small, minority denominations. This electoral arrangement meant that candidates from the various denominations figuring on a list had to co-exist, and to avoid extreme positions. Again, after 1937, it became established as a constitutional convention that the President of the Republic would be a Maronite, the Prime Minister a Sunni, and the President of the Chamber of Deputies a Shiite. Civil service posts likewise were distributed among the various denominations: thus, in 1946, out of a total of fifteen directors and directors-general in the administration, six were Maronites, three Greek Catholics, five Sunnis, and one Druse. Among the higher judiciary, two were Maronites and one was a Sunni; while of the holders of senior diplomatic posts, five were Maronites, two were Greek Catholics, three were Sunnis, and one a Shiite.

Notwithstanding the essential continuity between the *Règlement* of 1861 and the Constitution of 1926 in their approach to the governance of a multi-religious polity, the Lebanon of the Mandate was fundamentally different from the autonomous province of Mount Lebanon. The French, as has been seen, were entrusted with the Mandate in April 1920. On 1 September following, General Henri Gouraud, the High Commissioner, proclaimed the formation of a Lebanese state, far larger than the autonomous province—what came to be known as Greater Lebanon. This much larger territory represented a fulfilment of ambitions entertained by some Maronite writers and leaders, both ecclesiastical and civil, to create a Lebanese state which would regain its supposedly historical frontiers, and thus replicate the dominion of the emirate when it was at its largest extent. Greater Lebanon now included Beirut, Tripoli on the northern, and Sidon and Tyre on the southern littoral; it also took in the plain of the Biqa' in the east and the areas of Rachaya, Hasbaya, and Marja'yun in the south-east.

The evidence shows that this outcome resulted from pressure by Maronite leaders in Paris, from support of the Maronite cause by influential Frenchmen in France and in the Lebanon, and particularly by Gouraud who, like so many French officers, was imbued with Catholic sympathies. There was of course also the political calculation that a Greater Lebanon in the hands of the Maronites would remain a bastion of French power and influence, to act as a counterweight against Muslim turbulence in the neighbouring Syrian Mandate.

It was, as the sequel showed, an ill-advised policy for the Maronites to press for. At the time, Henri de Caix, the Secretary-General of the High Commission, who was perhaps the most acute intelligence to serve in the Levant during the whole French period, and who had designed the policy followed in Syria from 1920 until the Popular Front decided to do a deal with the National Bloc, expressed grave objections to the creation of a greater Lebanon. The Lebanon he argued for, before Gouraud took his final decision, would not have included Sidon and the areas to the south and east with their predominantly Shiite population. Tripoli in the north he believed should not form part of the new Lebanon state: it was 'a Sunni Muslim centre, somewhat fanatical, and in no way aspiring to be incorporated in a country with a Christian majority'. He also considered it a mistake to make Beirut the Lebanese capital. It was a large city, then predominantly Sunni, and, as he pointed out, in a few years' time it would have a population half the size of the whole Lebanon, the character of which it would greatly alter. In retrospect, de Caix had no doubt about the pressure which played its part in persuading Gouraud to establish Greater Lebanon. The pressure came from 'the old French elements in Beirut', reinforced by Christian communities affiliated with Rome—Maronites and Greek Catholics—who were, he wrote, too imbued with the *politique de clientèle* to see the dangers of the policy they favoured. Gouraud, de Caix said, was bombarded with talk about 'historical frontiers', the history of which no one took the trouble to establish. The Maronite Patriarchate, 'with that lack of genuine political sense which often characterizes cunning men with an aptitude for intrigue', for its part pressed Gouraud to the utmost to satisfy fully the ambitions of this French *clientèle*.

With its accretion of new territories and new communities, the Lebanese Republic became a much more difficult polity to govern. Michel Chiha, taking a hopeful view of its prospects, wrote that Lebanon was a country 'which tradition must defend against force'. The trouble was, however, that unlike the autonomous province of Mount Lebanon, the Lebanese Republic did not have the benefit of a coherent political tradition from which to seek guidance and inspiration. As has been seen, the Constitution of 1926 adopted the pattern established by the *Règlement* which itself had incorporated the older tradition of the emirates. However, the introduction of large numbers of Sunnis, who lived chiefly in the Beirut and the Tripoli areas, created from the outset an acute problem which was never resolved, and which in the end contributed greatly to the destruction of the Republic in the 1970s. These Sunnis found it very difficult to accept the idea of living as a minority in a state where the Maronites were dominant and also the cherished protégés of France. To start with, they would have nothing to do with the drafting of the constitution, and time and again, subsequently, they made clear their desire to secede from the Republic and to join Syria, which was, of course, overwhelmingly Sunni. The Greek Orthodox, likewise, felt little cordiality towards a state where the Maronites (associated with the Catholic Church, the traditional enemy, and clients of France, the traditional protector of Catholicism in the area) predominated. The Shiites in the south had been held in an inferior position by the Ottoman Sunni state. They were long habituated to passivity in politics; the French Mandatory had, furthermore, recognized them as a distinct community entitled to settle issues of personal status according to Shiite law, and not to Sunni law as hitherto. They were therefore not inclined to question the Mandate or Maronite predominance. All this, however, was to change radically from the 1960s, when Shiite demands and the manner of their prosecution were a main element in the destruction of the Lebanese Republic.

In the Lebanese Republic, then, the Maronites were to predominate. This was justified by their prominence in the emirate and in the autonomous province, but also by demography. A census taken in 1932 showed that Maronites living in the Lebanon numbered about 228,000, to whom might be added some 33,000 Maronites living abroad who retained Lebanese nationality and about 90,000 emigrants who no longer did, but who might still be considered to have retained a link

with the mother country. Sunnis living in the Lebanese Republic numbered about 178,000 and Shiites some 155,000. If one
added up all the Christian denominations whose members
actually lived in the territory or retained some connection with
it—an operation of questionable significance since not all
Christians had identical views or interests—the total amounted
to about 571,000. A similar—and similarly misleading—addition showed that the total of Muslims was about 361,000.

On the assumption that they were accurate, these figures
were used to justify a ratio of six Christian to five Muslim
deputies in the Chamber—a ratio which remained unchanged
up to the destruction of the Republic. Even at the time this
arrangement was made, it could be, and it was, questioned,
since the 90,000 Maronite emigrants—or the 160,000 Christian
emigrants—who no longer had Lebanese nationality, were used
to swell Maronite, and generally Christian, numbers to the
detriment of the Sunnis and Shiites, who had only about 20,000
emigrants to put into the balance. From the very beginning,
therefore, Maronite predominance in the Republic rested on a
very precarious foundation. Given that, in Professor Arnon
Soffer's words, in the Lebanon demography is the core of
politics and life, it is not surprising that the 1932 census was the
last one to be taken. There is, however, no doubt that if these
figures were accurate in 1932, half a century later they are no
longer so. From a careful survey and analysis of the evidence,
Professor Soffer has concluded that in 1983 the proportion of
Muslims (including the Druse) in the population amounted to
about 60 per cent, and of the Christians to about 40 per cent.
The proportion of the Maronites was about 24 per cent, of the
Sunnis 25 per cent, and of the Shiites 27.5 per cent.[18] These
figures mean that if and when the Lebanese Republic is
reconstituted, it would no longer be possible to assert Maronite,
and generally Christian, predominance.

Maronite leaders were indeed aware of the shaky position of
their community in Greater Lebanon, but they differed on
what had best be done to maintain its traditional predominance.
One of them, Emile Eddé, who served briefly as Prime Minister
in 1929–30, and as President of the Republic in 1936–41,

suggested in 1932 a radical policy of detaching both Tripoli and the south from the Lebanese Republic. Thus 140,000 Muslims could cease to add their considerable weight to the non-Christian part of the population, and the Christians would no longer fear the eventual disappearance of their paper-thin majority. His suggestion was not followed up, and the only guarantee of Maronite predominance, or perhaps even survival, remained the protection of France.

Another Maronite leader, Bishara al-Khuri, who was President of the Republic from 1943 to 1952, favoured another strategy—a bargain or an entente with the Sunnis, or more exactly with some of their leaders. His hour came after the expulsion of the Vichy French from the Levant in 1941. As has been seen, this was chiefly a British affair, and as in Syria, though the Free French formally succeeded their Vichy rivals in the administration of the Mandate, there was no doubt that the British were the dominant partner. Equally there was no doubt that the British favoured the satisfaction of what Churchill called 'Arab aspirations and susceptibilities'. Since the French were clearly in eclipse, those, like Eddé, who clung to the French connection, were likewise in eclipse. Khuri's strategy, swimming with the powerful current created by British policy, seemed much more promising in securing continued Maronite survival and even predominance. Khuri and his friends entered into an understanding with Sunni Moslem notables, among whom the principal figure was Riad al-Sulh—the understanding known as the National Pact of 1943. There was actually no document in which the terms of the Pact were spelled out and ratified. The Pact was more like an understanding between the Maronite and Sunni leaders about the policy to be followed in a fully independent Lebanon which would at once safeguard Maronite interests and make Sunnis feel at home in a state where Islam and Arabism did not reign supreme. The terms of this Pact or understanding were believed to run as follows:

1. Lebanon to be a completely independent sovereign state. The Christians to forego seeking protection or attempting to bring the

country under foreign control or influence. In return, Muslims to forego making any attempt to bring about any political union with Syria, or any form of Arab union.

2. Lebanon is a country with an Arab 'face' and language, and a part of the Arab world—having however, a special character.

3. Lebanon to co-operate with all the Arab states and to become a member of the Arab family, provided the Arab states recognize its independence and sovereignty within the existing boundaries.

The National Pact was a great departure in Maronite policies. It was a leap in the dark, fraught with uncertainties and dangers—particularly so because it accompanied yet another far-reaching change in the international context within which Lebanese politics operated. As has been seen, the British insisted that elections should be held in the Levant, and they finally were in the summer of 1943. Khuri and the grouping which he headed, the Constitutional Bloc, won a majority. This was not surprising since Eddé and his friends, organized in a National Bloc, were attached to the French cause, manifestly the losing side in the contest with Britain and its forceful representative, Sir Edward Spears. Khuri was elected President and Riad al-Sulh became Prime Minister.

When the Anglo-French forces invaded the Levant in 1941, the Free French proclaimed the independence of Syria and Lebanon. The proclamation, however, also spoke of a 'treaty in which our mutual relations will be defined'. The new Lebanese Government chose to interpret the 1941 proclamation to mean that the French Mandate had automatically lapsed, and introduced a bill into the Chamber removing all restrictions on Lebanese sovereignty contained in the 1926 Constitution—restrictions which the existence of the Mandate had made necessary. In spite of requests by the French Delegate-General who was then absent, for consideration of the bill to be postponed until after his return, the Chamber enacted the bill into law. On his return very shortly afterwards, the Delegate-General, considering this to be an act of defiance intended to humiliate the Mandatory Power, and furthermore that full independence had to wait on the negotiation of a Franco-Lebanese treaty, decided to arrest and intern the President and ministers. He did so, but then the British faced the Free French authorities with an ultimatum to the effect that either the President and the ministers were released and reinstated, or British forces would intervene. The Free French gave way. Khuri and the ministers were released and came back in triumph to Beirut. The French Mandate to all intents and purposes became meaningless.

1943, then, marked the beginning of a veritable new era for the Lebanese Republic. The Maronite leaders who won out in 1943 chose to steer it into the uncharted waters of Arab politics, into dependence on the goodwill of the Lebanese Sunnis and the neighbouring Muslim states. Simultaneously, enticed by British support, they rejected any French presence or influence in the Lebanon, and they were entirely on their own in steering Lebanon through the treacherous eddies of inter-Arab politics. It may be seen in retrospect that the autonomous province was a success, and that Greater Lebanon could function tolerably well, in the one case because the Powers exercised an oversight over its affairs; and in the other because the Mandatory was, so to speak, a court of last resort standing above, regulating, and moderating communal rivalries and conflicts. With the disappearance of such outside restraint, the Lebanese polity was like a prestidigitator, keeping a large number of balls in the air, seemingly without the use of hands. It could not, and certainly did not, last.

In the Lebanese electoral system, voting in constituencies, as has been seen, was by lists. These lists were made up by the notables belonging to the largest community in each constituency, who co-opted candidates from the various communities which the law specified for the particular constituency. This meant that there had to be give and take between the representatives of the various communities, that political moderation was a vital necessity. It also meant that candidates on these lists would have to be able to confer favours on followers in their respective communities, to procure official posts, to facilitate official procedure, and the like. The other side of this coin was, of course, corruption more or less extensive, more or less blatant, enmeshing in its network the small man and the big notable alike. It stands to reason that the main fount of corruption is the administration itself, above all its head, the President, who is able to grant or withhold favours, and when necessary judiciously to shut an eye. It may well be that corruption is inescapable, indeed necessary, in a polity made up of various groups held together by little except the necessities of co-existence.

Khuri, elected President in 1943, was to serve, according to the constitution, for a single, non-renewable six-year term. In 1947, two years before the end of Khuri's mandate, parliamentary elections were held, as due. The elections seem to have been rigged by the President and his followers, and a pro-Khuri majority was elected. In 1948 this majority voted to suspend in Khuri's favour the non-renewal clause. The rigged elections and the prolongation of Khuri's term of office made his fellow notables and competitors angry and disgruntled, the more so that an election in 1951 was equally rigged in Khuri's favour. Since there was no superior authority in the shape of a mandatory or the like to appeal to, and since the Chamber of Deputies was now Khuri's creature, the President's opponents carried their opposition to the streets, seeking to exploit economic discontent, and mobilize those who felt that the regime was indifferent to their interests and partial in the distribution of its favours. Khuri's position was weakened by the murder of his principal Sunni ally, Riad al-Sulh, in 1951. Maronite rivals, chief of whom was Camille Chamoun, banded together with a powerful Druse leader, Kamal Jumblat, and with Sunni and Greek Orthodox opponents of the regime, in order to bring it down. In the summer of 1952, there occurred the *coup d'état* which brought the monarchy to an end in Egypt. The coup was hailed in the Arab world as a blow against reaction and corruption. The atmosphere created by it was favourable to an opposition claiming to work for the eradication of all corruption. The opposition succeeded in organizing popular agitation and in calling, during September, an effective general strike. The President did not have a reliable Sunni ally willing to become prime minister, and the commander of the army was not prepared to take action against the agitation and the strikes. Khuri gave in to the tumult, and resigned. Khuri's Parliament elected as President his chief Maronite opponent, Camille Chamoun. These events showed that constitutional order had broken down, and that the careful and complicated constitutional and electoral arrangements gradually developed from 1926 onwards were by no means able to contain and resolve political disputes.

Two elections took place during Chamoun's tenure, in 1953 and 1957. Both were preceded by changes in the electoral law. For that of 1953, Chamoun abolished the five large constituencies into which the country was divided, and which went with the list system. He replaced them with thirty-three constituencies of which twenty-two were represented by a single member. The effect of this was to curtail the power of the big notables who headed the competing lists under the previous dispensation, by curtailing their patronage and the followers who were beholden to them. The purpose seems to have been to destroy Khuri's Constitutional Bloc as an electoral force, and in this Chamoun was successful. Having in 1953 reduced the number of deputies from seventy-seven to forty-four, in 1957 Chamoun increased it to sixty-six, and the constituencies were again enlarged and reduced from thirty-three to twenty-five. The outcome of Chamoun's gerrymander—analogous to Khuri's gerrymander—was to pack the Chamber with his clients and followers.

The same tension which had arisen during Khuri's last year was also now evident for much the same reasons. Chamoun's rivals and opponents were disgruntled as he himself had been when Khuri was in power. In 1957–8, which would have been the last year of Chamoun's tenure under the Constitution, there was widespread suspicion that, like Khuri, he would seek a renewal. Chamoun refused to still these suspicions, declaring at the end of December 1957 that, though he was in principle opposed to an amendment of the Constitution, yet he would have to reconsider his position if there was no one to succeed him who would ensure continuity for his own policies.

In 1958, however, other events impinged which made the tension much more acute than it had been in 1952. In February,

Syria decided to join Egypt in a United Arab Republic under Nasser. The union evoked great enthusiasm in the Arab world. The Lebanese Sunnis fully shared this enthusiasm, and began to entertain the prospect of being included in this new powerful Arab state, under the leadership of someone who was widely worshipped as the saviour of Arabism. This, of course, was to go clean against, in fact it was virtually to renounce, the idea of the 1943 National Pact. Chamoun himself, who ironically had been a prominent supporter of the Pact and of Khuri's policy in 1943, was now considerably cooler towards the Arab connection. During the Suez affair in 1956, he failed to embrace Nasser's cause as fervently as his Sunni compatriots desired, and subsequently accepted the so-called Eisenhower Doctrine which sought to limit the influence in the Middle East of Nasser's Soviet friends, and which Nasser and his followers in the Arab world categorically anathematized. Agitation organized by Chamoun's opponents, and fed by the United Arab Republic, rose to a crescendo in the first months of 1958. Very soon clashes between Chamoun's supporters and his opponents degenerated into a civil war between and within communities, in which Chamoun's opponents were sent arms, volunteers, and money from Damascus, while Chamoun's supporters benefited from the help of the monarchical regime in Iraq, which was at daggers drawn with Nasser. The troubles began in May, and soon large parts of the country, mainly Druse and Muslim, were out of the control of the Government. Lebanon complained to the United Nations Security Council. This did not advance matters one whit. A United Nations Observer Group was sent to monitor the traffic of arms between Syria and Lebanon. The Group was much too small for the purpose and did not try to find out what went on at the Lebanese-Syrian border at night. Anyway, Hammarskjöld, the United Nations Secretary-General, had a soft spot for Nasser and in public denied that the United Arab Republic was much—or at all— implicated in the civil war. In private, however, he told Nasser that he had overplayed his hand, that military intervention and incitement to rebellion through the radio had to stop.

Hammarskjöld's expostulations proved useless. The disorders went on, fanned by the United Arab Republic. There is no telling where or how they would have ended, had not the military *coup d'état* in Baghdad destroyed the Iraqi monarchy on 14 July. This created widespread fears that other regimes friendly to the West would also soon fall to Nasser's subversion, and that behind him stood the Soviet Union. The US decided immediately to send the Sixth Fleet with a large contingent of troops who landed in Beirut. The disturbances soon died down. Chamoun's position was greatly weakened by the civil war. It was out of the question now to ask the Chamber of Deputies to override the provisions of the Constitution. With powerful Druse and Muslim leaders, controlling bodies of armed men, ranged against him, a majority in the Chamber, which he could probably have obtained, would have been meaningless. His position had also been considerably weakened by the adamant refusal of the Army Commander, General Shihab, to order his forces to act against the armed bands which were challenging the lawful government. Shihab argued that if the Army became involved in communal disputes, it would disintegrate. This is in fact what happened in the civil war which began in 1975; the Army fell apart, each contingent siding with its own community. Shihab's stance on this occasion, which recalled his earlier one when he refused to support President Khuri in 1952, tellingly indicates that independent Greater Lebanon steering itself by the National Pact could not be a going concern. The civil war of 1958 was thus an event which by its very violence strips away illusions which, had they not been stripped away, might have continued, for a time at any rate, to act as a preservative for the body politic.

At the end of July, Chamoun's Chamber met and elected Shihab as President. His attitude during the civil war no doubt made him acceptable to the Muslim and Druse leaders, as well as to Nasser. The US seems also to have approved of the choice, as one likely to contribute to the appeasement of communal conflicts which, when the initial moment of panic passed, they no longer saw as opening the way to Soviet hegemony in the Levant. To judge by his policies during the six years of his presidency, Shihab had concluded that Lebanon should cease to be a collection of communities dividing among themselves powers and privileges, and should become a modern state in which the government is the guardian of the general interest, and the agent of economic and social modernization. This meant that Shihab would try to abandon as much as possible the kind of politics hitherto practised, which revolved round coalitions of powerful communal and regional notables. Shihab tried to govern through a bureaucracy renovated and directed by technocrats, and through the *Deuxième Bureau*, i.e. the Army Intelligence Directorate. His style of government was, in Professor Kamal Salibi's words, 'secretive, authoritarian, and to some extent underhanded'. It elicited the usually passive opposition of the long-established vested interests, and more particularly of the Maronites. This was because Shihab's activist style meant that the state, in trying to be even-handed as between the various communities, tended to favour the Muslims and in particular the Shiites, as being the less advantaged part of the population. Another cause of Maronite discontent was that Shihab took care to be friendly to Nasser, before and after the dissolution of the United Arab Republic in 1961. He may have judged such caution to be necessary for the survival of Lebanon, but following the turmoil of 1958, and of Nasser's role in it, it is not surprising that Maronites did not view it with favour.

Shihab embarked on an active welfare policy, providing running water, electricity, and medical dispensaries in remote areas, extending roads and improving the government school system. But his policies and actions during his six-year term of office could not by any means change the fundamental character of the polity, or in any appreciable degree diminish the power of long-established notables. The reforms, however, hastened, if they did not initiate, social change which was shortly afterwards profoundly to disturb Lebanese politics, and eventually to contribute to the ruin of the Lebanon. As has been seen, de Caix forecast in 1920 that Beirut would in due course contain half the population of the Lebanon. In the 1960s migration to Beirut and other cities from the countryside for various reasons gathered increasing momentum. The new migrants changed the character of urban areas, and this in turn itself changed the character of Lebanese politics. The Shiites, in particular, migrated in large numbers from the South and formed an impoverished and volatile element to add, in particular, to that represented by the Palestine refugees, who had come in the years after the Arab–Israel wars of 1948–9, and who had settled in camps which, with the years, became permanent.

It was in those years that the Shiite mass, long a passive element in Lebanese politics, acquiescing in the leadership of traditional notables, began to be restless and quickly learned to practise an activist style of politics. This was very much the doing of Musa al-Sadr, a Persian cleric of Lebanese origins who came to the Lebanon in the late 1950s and very quickly assumed a very prominent position in the Shiite community. After the formation of the Palestine Liberation Organization in 1964, commando raids started to be mounted from the Lebanon against Israeli territory, and the Israelis retaliated with armed incursions into the Lebanese South. These armed clashes necessarily affected increasing numbers of Shiite villagers who fled to Beirut from the fighting. Sadr took up the cause of the Shiite masses in the South, railing against the general and long-standing neglect of the area by the Government, against the corruption and greed of the establishment, and demanding military protection against the Israeli raids. He organized a Shiite strike in Beirut to drive home his protests, and set up a Movement of the Deprived, and gave it a military arm, the Hope (*Amal*) Movement to provide protection and defence for his community. His rhetoric was radical, belligerent, and rousing: 'We do not want sentiments, but action . . . from today on I will not keep silent . . . we want our full rights completely . . . O rising generations, if our demands are not met, we will set about taking them by force: if this country is not given, it must be taken'.[19]

As clashes between Israel and the PLO in southern Lebanon worsened, affecting increasing numbers of Shiites, Sadr increased his agitation, and his attempts to organize bodies of armed men to defend the South against the effects of the

Israel–PLO conflict: arms, his slogan went, were an ornament to men. But arms ostensibly to guard against Israeli incursions could obviously also be put to other uses. The other theme which figures in Sadr's oratory was that of struggle against the injustice of which Shiites in the Lebanon were victim, and struggle for a new Lebanon where equality between all citizens would obtain.

In his campaign on behalf of the Lebanese Shiites, Sadr established connections with other Arab governments, namely Syria and Libya. Syria was ruled by Ba'thist officers who were predominantly Alawite, and Alawites, as has been seen, were considered heretical by Muslims. Sadr recognized the Alawi sect as forming part of Shiism, thus conferring on it a sort of respectability in order presumably to gain the support of the Syrian rulers. He also maintained a connection, shadowy and obscure, with the ruler of Libya, Colonel Qadhafi. This connection was to lead to his disappearance, and presumably his death. In August 1978 Sadr and two of his associates went to Libya, it is said at Qadhafi's invitation, and all three disappeared, so far without trace, after their arrival in the country. Sadr's attempts to use his connections with other states in order to strengthen his hand against the Lebanese Government, is on a par with similar Lebanese Sunni actions during and after the 1958 civil war. It was yet another sign that the National Pact, the pillar of Lebanese independence since 1943, was on the point of crashing to the ground.

Shiite activism, as encouraged and directed by Sadr, was to be manifest from the late 1960s onward, some years, that is, after Shihab's presidency. Shihab, it was thought, wanted to do the same as Khuri and Chamoun, namely to amend the Constitution and be elected for another term. However, there was a great deal of opposition to such a move from a coalition of notables, Maronite and Sunni, supported by the Maronite Patriarch. Shihab gave up the idea. To succeed him, the Chamber elected Charles Helou, journalist, diplomat, and minister. He had been a supporter of Khuri's Constitutional Bloc and hence of the National Pact. The policy of the Pact was also followed, not to say emphasized, by Shihab who

allowed Nasser, whether wittingly or not, to interfere in
Lebanese internal affairs so as to ensure its alignment with
Egyptian policies. Helou, who had no great political standing,
found himself in the hands of Shihab's followers in the administration and in Army Intelligence. He was, therefore, derided
by his rivals, and by those who looked askance at the Shihabist
grip on the levers of government. Helou, wanting to establish
an independent position, encouraged the anti-Shihabist notables, and Shihab's entrenched followers in turn sought to create
difficulties for him.

Such games, however, and the talents required to play them,
were fast becoming irrelevant to the grave problems which now
began to confront the Lebanese Government. In 1964 the
Palestine Liberation Organization had been formed, at Nasser's
instance. Lebanon, with a large number of Palestinian refugees,
and on the northern borders of Israel, at once became a centre
of PLO activity directed against Israel. As has been said, Israel
reacted by attacking the areas of south Lebanon where the
PLO guerrillas were based. It was not only the Shiites who
were affected by such developments. The Sunnis looked upon
the Palestinian cause as their own, and were unwilling that
PLO activities against Israel originating in the Lebanon should
be curbed. In 1968, a Lebanese Muslim who took part in a
PLO action against Israel was killed. At his funeral afterwards
there were large demonstrations calling for Palestinian guerrillas to be given freedom of action on Lebanese territory. At one
of these the Prime Minister, Abdullah al-Yafi, declared himself
to be in favour of this. This could not but alarm the Maronites.
Elections were taking place that year, and the Palestinian issue,
with all its internal repercussions, could not fail to be widely
agitated. In these elections three Maronite political groupings:
Chamoun's followers, those of Raymond Eddé (a son of
Emile's), and Pierre Gemayel's Phalange Party banded
together, with the support of the Maronite Patriarch, in order
to express their alarm at the threat to Lebanon's internal
stability and its external security posed by the PLO and its
Lebanese supporters. This coalition, known as the Triple
Alliance, proved very successful in the elections, and this would

indicate that the fears of its leaders were shared widely, at any rate among the Maronite electorate.

PLO operations against Israel—and Israeli retaliation—continued apace. One spectacular Israeli response was to attack Beirut International Airport on 31 December 1968, destroying thirteen airliners which were on the runways. There were loud protests by PLO supporters who asked why the Lebanese Army had taken no action against the Israeli invaders. Students in Beirut struck and demonstrations against the regime followed, which led to the fall of Yafi's Government.

Yafi's successor, Rashid Karami, was no more successful in dealing with the PLO issue. In April 1969 the Lebanese Army clashed with demonstrators in Sidon, a predominantly Sunni city, who were protesting against its attempts to curb and control the PLO. The gap between the Maronite and the Sunni notables over the PLO issue went on widening, even though Karami tried to reach a compromise whereby the Lebanese Army and the Palestinian guerrillas would 'co-ordinate' their activities. Clashes between the Army and the armed Palestinians continued, and they once again brought overt interference by other Arab states in Lebanese affairs. Under their pressure, Helou agreed to try Karami's idea of 'co-ordination', empty and meaningless as it was, and a Lebanese Army delegation left for Cairo towards the end of October 1969, in order to specify the details and the workings of this co-ordination. They met Yasir Arafat, the Chairman of the PLO, in the presence of the Egyptian Ministers of War and of Foreign Affairs, and an agreement was worked out and signed by the Lebanese with the PLO on 3 November. By its terms, the Lebanese allowed the Palestinians to establish posts and armed units in Palestinian camps, which came under the jurisdiction of the PLO. Palestinians could freely join the guerrillas, whose movements to the Israeli border were to be facilitated by the Lebanese authorities, which would also guarantee the guerrillas' lines of supply. There were to be fixed bases for them in south Lebanon. The PLO, in return, would ensure that there would be proper discipline within its ranks, and that PLO forces would not interfere in Lebanese politics.

The Cairo Agreement was denounced by the Maronites as a grievous betrayal of Lebanese sovereignty. Nor did it serve to diminish Palestinian interference in Lebanese politics. The PLO could not forbear doing so on the calculation that if their Lebanese supporters were strengthened, they would be able to have greater freedom of movement in their war against Israel. In any case, after September 1970, the ranks of the guerrillas were swelled by their brethren who had fled from Jordan, after the defeat of their armed challenge to King Husayn. The Sunnis were moved by loyalty to co-religionists and by the hope that a stronger PLO in the Lebanon would mean an accession of strength to them. In the spring of 1973, the Israelis raided Beirut and killed a number of PLO leaders. PLO violence against the Lebanese authorities erupted. In one clash, PLO commandos abducted three Lebanese soldiers. Their release was demanded and the Army took up positions around Palestinian camps in the south of Beirut. Clashes continued not only in Beirut, but elsewhere in the Lebanon where guerrillas penetrated from Syria to attack Lebanese army positions. Syria closed its frontiers to Lebanese traffic and denounced Lebanese complicity with foreign attempts to liquidate the Palestinian struggle. Clashes between the two sides went on until 18 May, when the Army and the Palestinians signed an agreement at the Malkert Hotel in Beirut designed to reaffirm and clarify the Cairo Agreement. The Malkert Agreement, however, proved as useless as its predecessor. The more so that the October 1973 war and its sequels raised tensions still further, by increasing the PLO's standing and self-confidence, and exacerbating Maronite fears.

The rising tension and the manifest inability of the Lebanese army to cope with it finally brought on to the scene, in a most prominent manner, armed Christian and Maronite popular militias who increasingly took upon themselves the defence both of the communities from which they sprang, and the Lebanese state which some of them viewed as a Christian refuge, and others as a polity composed of different co-existing communities. One of these armed groups was composed of the followers of Camille Chamoun, but the largest and most influential was that of the Phalangist Party, founded and led by Pierre Gemayel. Originally known as the Phalanges Libanaises, it was founded in 1936 by five Lebanese Christians, four of them Maronites. The founders were by no means the notables around whom Lebanese politics then and later revolved. Gemayel was a pharmacist, and his co-founders were two journalists, an engineer, and a lawyer.

Both the manner and the character of the original group were clearly influenced by political tendencies very much prominent in Europe in the mid-1930s and much imitated in the Middle East: the tendency to view politics as akin to armed struggle, and the tendency to pursue an ideological style of politics. At their foundation the Phalanges did not have more than 300 members. But their appeal grew very quickly, and in 1943 they numbered over 38,000. In that year they took the part of Khuri and his Government in their defiance of the Mandatory. This gave them a new position of great respectability in Lebanese politics, and with the increase in tension in the country created by the Palestinian armed struggle, they came to be seen by large numbers of Maronites as their shield against the Palestinian guerrillas and their Lebanese Muslim sympathizers. By 1968 the Phalangist Party had over 64,000 members, 80 per cent of whom were Maronites, and 10 per cent other Christians. While the Party went on taking part in parliamentary and electoral politics, and propounding policies for the reformation and modernization of the Lebanese state—policies far removed from the ideas and attitudes of the traditional notables—they were simultaneously arming and training their members in anticipation of the clash with the Palestinians and the Muslims which they believed was bound to come.

And come it did. On the morning of 13 April 1975 Gemayel arrived in a Christian suburb of Beirut for the consecration of a church. While the ceremony was going on, the armed Phalangists who were guarding the approaches of the church stopped a car with a covered licence plate and turned it back. Another car also with a covered licence plate followed. It went through the Phalangist cordon, its occupants firing in the direction of the church and killing four men, including one

Gemayel bodyguard and two members of the Phalangist militia. In retaliation, a bus with Palestinian passengers returning from a parade of commandos was ambushed in the same suburb and all the passengers were shot dead. This inaugurated an endless cycle of violence in which all groups and communities in the Lebanon sooner or later became involved. Whole neighbourhoods and cities, including Beirut, were gradually devastated, looted, and destroyed. Life became cheaper than the cost of a bullet or a grenade. It became more and more dangerous for one community to inhabit the same neighbourhood as another, inimical, community. The confessional map of Beirut and of the Lebanon changed drastically, the members of each community huddling for safety with their co-religionists. In the striking expression coined by a geographer, Dr N. Kliot, the geography of Lebanon became a geography of hostages—every community hostage to every other.[20] These events quickly destroyed constitutional government as it had existed from the beginning of the French Mandate. It is not only militias which stand confronting one another in gladiatorial posture. Foreign armies, also, and the governments controlling them, have also played their part in the dissolution of the Lebanese state. Syria intervened quite forcibly early in the conflict, first through Palestinian militias which it controlled, and later on directly through its own army. Syria came to control large parts of eastern and northern Lebanon and of Beirut itself. Israel, too, considered that vital interests required its intervention in the Lebanon—an intervention which culminated in an invasion in June 1982, and in the subsequent control of a sizeable portion of south Lebanon.

Under these circumstances, parliamentary elections could not take place, the Chamber could not meet, courts could not function, and the administration was under the control of whichever militia or army held sway over territory in which it was supposed to operate. There continued, however, to be a Lebanese Government leading a ghostly existence, complete with a President, Prime Minister, and ministers, and a network of embassies throughout the world. In one of the sequels to the Iraqi invasion of Kuwait in 1990, Syria was at last able, through the complaisance of the US—anxious to recruit Arab allies against Iraq—to impose itself in the Lebanon and make it into a satellite.

5. Politics in Iran under the Pahlavis

As was seen in Chapter 3, the political situation in Persia became disturbed and chaotic after the outbreak of war in 1914. Persia was a cockpit in which the British, the Russians, the Ottomans, and their German allies vied for influence and control. The Russians had troops in the north, the British in the south, and Ottoman forces occupied parts of western Persia. Conditions became even more disturbed after the outbreak of the Russian Revolution in 1917. A guerilla movement known as the Jangalis succeeded in controlling much of the area of Gilan, in northern Persia. In Azerbaijan, to the south-west of Gilan, Azeri radicals who had been prominent in the Revolution of 1906, demanded autonomy for the province and such measures as land distribution. The British were afraid that the collapse of the Russian forces in northern Iran, which had followed the Bolshevik revolution, would allow the Ottomans to attack the Caucasus through Persian territory. At the end of the war, they became concerned that Bolshevism would obtain a foothold in Persia and threaten the substantial British strategic and political interests there. They therefore sent various bodies of troops to northern and eastern Persia. However, with the end of the war, there was pressure from within the British Government in London, and from the Government of India, for these troops to be withdrawn. At the end of 1920, British officers and officials however were afraid that 'Bolshevik Committees' would take over the Persian Government. They were also concerned about insubordination and lawlessness said to be rife among Persian troops in Tehran.

It thus seemed to the British commander of the North Persia Force, Major-General Ironside, a good idea to encourage and help Colonel Reza Khan, the forty-two year-old commander of the Cossack Brigade—the most efficient formation in the Persian army—to march with his troops from Qazvin south to the capital, re-establish order and make sure that there would be no Bolshevik takeover. On the night of 21 February 1921, Reza Khan entered Tehran with his 3,000 troops, arrested some sixty public figures, imposed his nominee, a pro-British journalist, as prime minister, and assured the Shah that his sole purpose was to save the monarchy from revolution. The Shah elevated Reza Khan to the newly-created post of Army Commander. The following May he became Minister of War, and in October 1923 Prime Minister. He was then granted the title of Commander-in-Chief—a title which the Constitution had vested in the Shah. Finally, in December 1925, the Qajar dynasty was deposed by the Parliament, and the state was entrusted to Reza Khan. Thus, at the age of forty-seven, this officer with obscure antecedents had risen from the ranks and become Reza Shah. He adopted for his family the ancient Persian name of Pahlavi and, like Napoleon, himself put the crown on his own head.

Reza Khan established his power after 1921 by considerably enlarging the armed forces and subduing the Jangalis and tribal separatists among the Kurdish, Azeri, and Persian tribes, the Baluchis, and the Turkomans, as well as the Arab tribes of south-west Persia and their leader Sheikh Khaz°al of Mohammerah. Reza Khan also used the army to manipulate elections and to have those who were ready to do his bidding returned. Universal male adult suffrage had been enacted by the third National Assembly in the chaotic conditions of Persia after the outbreak of the world war, pushed through by the Democrats, i.e. the radical constitutionalists who had finally triumphed over Muhammad Ali Shah. The consequences of this reform were described many years later by a well-known poet and Democrat, Bahar (b.1886) whose life spanned the constitutional revolution, the advent of Reza Shah, and of his son and successor, Muhammad Reza Shah.

This electoral law, which continues to plague the country even today in 1944, is one of the most harmful and least thought-out bills ever passed by us Democrats. By introducing a democratic law from modern Europe into the paternalistic environment of traditional Iran, it weakened the liberal candidates and instead strengthened the conservative rural magnates who can herd their peasants, tribesmen, and other retainers into the voting polls.[21]

Revealing as this comment is, particularly when coming from a Democrat, it is yet misleading in its contrast between the electoral fortunes of 'liberal candidates' and of 'conservative rural magnates'. Before and after Reza Khan's enthronement, 'liberals' were rather enthusiastic in their support. They saw him as the reformer and modernizer which Persia needed. It was under Reza's auspices and through his patronage that the Revival Party, heir to the pre-war Democrats, was set up and gained seats in the National Assembly. These 'liberals' and progressives worked faithfully for the new regime. Davar, a jurist, reorganized the Ministry of Justice, introducing laws based on the French Civil Code and the Italian Penal Code; he eventually committed suicide fearing disgrace or murder. Teymourtash, a progressive landowner who supported Reza Khan from early days and served as his court minister after 1926, was put in prison in 1933 on charges of bribery, extortion, and embezzlement, and died five years later, allegedly of a heart attack. A Qajar prince who served as Reza's right-hand man after 1923 was dismissed in 1930 for misappropriating state funds and strangled to death eight years later while under house arrest. Tadayon, a leader of the Revival Party and Minister of Education, went from office to jail for complaining that the budget of his Ministry was too low and that of the Ministry of War too high. Other liberals or progressives who served Reza were luckier in being simply dismissed or allowed to retire peacefully when they lost the monarch's favour.

As for the 'conservative rural magnates' of whom Bahar also spoke, they certainly did not fare better than the liberals. As'ad Khan, the Bakhtiari chieftain, who supported Reza in his campaign against other tribes after 1923, was dismissed as War Minister, jailed without trial, and murdered in prison. Abdul Hasan Diba, wealthy landlord and assistant minister, uncle of Empress Farah, Muhammad Reza Shah's third consort, was dismissed, and murdered while awaiting trial. Reza Shah systematically dispossessed and subjugated the landowning classes, as much to create a centralized State as to enrich himself. When he abdicated he had in his possession 3,000,000 acres, mainly in the fertile province of Mazanderan from which he hailed. The Shah diminished and treated with contempt the men of religion. The legal reforms which Davar had introduced did away, in large measure, with their judicial functions. In 1939 the Shah decreed that lands held by religious foundations should be taken over by the state.

Reza's centralizing policies, his hostility to the religious establishment, and his determination to modernize Persia recall Atatürk's similar endeavours in Turkey. Atatürk, however, inherited a state which had been practising modernization and centralization much more thoroughly and much longer than Persia. Atatürk, again, was not a man of blood who killed both his servants and his opponents without scruple, nor did Atatürk expropriate to his own benefit the possessions of his subjects. Persia, again, unlike Turkey, was, when Reza Khan took over, criss-crossed with localisms and separatisms; it also had a religious establishment which was much less dependent on the state than the Sunni divines of the Ottoman Empire. Reza uprooted and destroyed all this and became, in effect, the sole and absolute ruler of his domain. Elections continued to take place, and the National Assembly continued to meet. The elections, however, proceeded according to the Shah's bidding, and the National Assemblies likewise. As early as 1927, one year after Reza Shah's enthronement, the British Minister in Tehran wrote that

the Persian Majles [National Assembly] cannot be taken seriously. The deputies are not free agents, any more than the elections to the Majles are free. When the Shah wants a measure, it is passed. When he is opposed, it is withdrawn. When he is indifferent, a great deal of aimless discussion takes place.[22]

This remained the case all through Reza Shah's reign, which ended with his abdication in September 1941. The previous August Iran was invaded from the north by Soviet troops and from the west by British troops. The two allied Powers were afraid of German activities in Iran (the name which the Shah decided to substitute for Persia in 1935, to emphasize the country's ancient Aryan heritage, which Nazi ideology had made fashionable) since Reza Shah had sought German friendship as a counterpoise to the two Powers who were in his neighbourhood, Great Britain and the Soviet Union. Also, Iran became a highway for military supplies to reach the Soviet Union from its allies, and its security, therefore, a vital interest.

The army, the object of Reza Shah's unremitting care over the years, proved useless in resisting the invasion. His position, feared and hated as he was, became impossible, and his abdication was inescapable. He was sent into exile in South Africa where he died in 1944. These events administered a great shock to the monarchy and to the pervasive system of control which Reza Shah had set up. His son, Muhammad Reza Shah, had perforce to abate considerably the pretensions of the court and to adopt a moderate and conciliatory stance. He was able to retain control over the army, but had to allow the National Assembly a much greater say in legislation, in debate, and in control over ministers. The freedom of speech in the Assembly which the change of ruler suddenly made possible, may be exemplified by the admonition which a member addressed to the new Shah soon after he had ascended the throne. This member, Ali Dashti, had edited a newspaper which from 1922 onwards supported Reza Shah. He was rewarded by a seat in the Assembly, but he subsequently lost favour, his parliamentary immunity was lifted, and he was confined to a sanatorium. He was afterwards released and he resumed his seat in the Assembly, from which he now felt free to warn the son of his patron that if he meddled in politics he would lose his throne.

The new freedom to speak and to challenge the government was made possible not only by the disappearance of a despotic shah, but also by the foreign occupation. The occupants were two Great Powers who had had a long history of rivalry in the country. Even though one of them, occupying the north, was no longer Tsarist Russia but the Soviet Union, yet the very fact of occupation attracted clients to it who were presented with tempting opportunities. As for Britain, it had not suffered the same eclipse which its rival had after 1917, but maintained in the south a network of clients and supporters. The rivalry of these two Powers was now most clearly manifest at election time. Thus in the elections for the Assembly which took place during November 1943–February 1944, we find that, according to the Foreign Office in London, Soviet army lorries in Azerbaijan were used to transport workers from factories to voting booths, in order to nullify the attempts by the Governor-General to prevent them from voting by closing the poll at six o'clock—the time when factories closed—and by positioning the voting booths outside working-class neighbourhoods. The Foreign Office commented that they were hardly in a position to show indignation since British consuls themselves vetoed unacceptable candidates in their own zone of occupation.[23]

The presence of these two rival Powers had its repercussions not only on elections, but also on the functioning of the Assembly. It did not take long after the occupation for rival factions to be formed, and factional disputes to proliferate. Some deputies favoured the British and some the Soviet connection; some supported the court, and some wished to pursue liberal and progressive policies. These last became attracted to the United States following its entry into the war, and its interest in sending mlitary supplies to the Soviets through Iran. In their eyes the US served as a counterpoise to Britain and the USSR. These factions were by no means stable and a great many were the creature of a day: they formed and re-formed, and their numbers grew or shrank in response to manœuvrings, bargaining, and intrigue. Thus the 1943–4 elections started with sixteen parties and ended with forty-two, but these numbers shrank considerably in the ensuing Assembly.

The Assembly was now free to choose the Prime Minister. The choice was, of course, in the final resort subject to the wishes of the occupying Powers whose priority was the security and good order of the country, but who each also tried to ensure that the administration would be friendly to them and not to their rival. An Iranian Prime Minister was now no longer the obedient servant of a single, imperious master. He depended rather on changing and ephemeral majorities in the

Assembly, he had to satisfy the exigencies of the occupying Powers, and he had to contend with the popular discontents created by the wartime scarcity of food and other essentials, and to the inflation created by the local spending of the Allied armies. He also had to reckon with the Shah who still had great influence and many followers, and who still controlled the army. It could be a daunting task, yet there were always willing candidates, tempted by the glamour of office, the prospect of power, and possibly also by a potent desire to get the better of a son whose father had treated them with contempt or brutality. One of the most prominent of these was Ahmad Qavam (1878–1956) who came from an aristocratic family and who had begun his career as private secretary to Muzaffar al-Din Shah. He had supported Reza Khan who in effect made him Prime Minister in 1921. Reza, however, later accused Qavam of plotting against him and he had to flee to Europe. In 1942 he received enough support to be nominated Prime Minister and he assumed office in a tense situation. There were food riots in some provincial towns and great restiveness among the tribes.

On taking office, Qavam declared that henceforth the chief of staff would no longer take orders from the Shah, but would be subordinate to the war minister whose portfolio he himself kept. He purged the war ministry of the Shah's dependants and resolved that his ministers would henceforth communicate with the Shah only through him. A trial of force between Shah and Prime Minister ensued. In December 1942 a demonstration to protest against inflation, the high price of food, and a bill imposing income tax (drafted by an American adviser appointed by Qavam) was organized outside the Assembly by the bazaar guilds. The demonstration turned into a riot when hired thugs attacked the Assembly, and the army prevented the police from intervening. The rioters beat up two deputies, looted bakeries, and other shops, as well as Qavam's residence. Allied troops had to intervene. Shortly afterwards Qavam resigned.

Governmental instability, then, characterized Iranian politics after 1941. It was the outcome of struggle between a young Shah who did not have his father's commanding position, of rivalries between the Great Powers who, so long as the war lasted, had the last word, and of the interplay between political factions in the Assembly—changeable, quarrelsome, and plunged in a welter of endless intrigues, the point of which outsiders certainly, and they themselves most probably, had difficulty in comprehending. This instability is exemplified by the fact that in the lifetime of the Assembly which sat between 1944 and 1947 seven Prime Ministers formed nine administrations in which 110 ministers held office.

Of the Powers which held sway over Iran, the Soviet Union pursued the most activist policies. Its prestige and influence enabled Iranian Communists to come out into the open, and to organize political parties and trade unions. Thirteen days after Reza Shah's abdication, twenty-seven Communists who had been imprisoned in 1937 met and formed the Party of the Iranian Masses, commonly known by its Persian name, Tudeh. Soviet presence in the north and British acquiescence in the south because of the wartime alliance, allowed them to recruit members and to organize workers in trade unions, the activities of which were as much political as industrial. So long as the Shah and the politicians jockeyed for power, some politicians would find the Party of the Masses a useful weapon in their struggle. In the northern provinces under their occupation, the Soviets encouraged Azeri and Kurdish separatism. In Azerbaijan, a former Comintern agent, Ja'far Pishevari, established a Democratic Party of Azerbaijan and prepared for an armed uprising by means of weapons passed on to them by the Soviet army, which had been seized from the Iranians in 1941. In November 1945 Pishevari occupied the Government offices in Tabriz and cut communications with Tehran, while Soviet troops prevented the Iranian army from entering the province. In Kurdistan, likewise under Soviet occupation, a Kurdish People's Republic was proclaimed in December 1945.

Azeri and Kurdish attempts at secession, encouraged by the Soviets, meant a serious crisis for the government in Tehran. Since the Soviets were also demanding oil concessions in the north, to match those which the British had in the south, it was clear that they were trying to establish paramount influence in the country. It was, on the other hand, not clear then that Iran could count on either Britain or the US to checkmate the Soviets, and it would thus have to deal with them on its own. The Soviets, whose actions were unfriendly and even hostile, indicated that they preferred to negotiate with Qavam, who thus became Prime Minister in March 1946. He remained in office until November 1947. He conducted very skilful negotiations with the Soviets, and succeeded—with US and British help—in persuading them to evacuate northern Iran, which meant the end of Azeri and Kurdish separatism. During his tenure Qavam resumed his earlier attempt to diminish the Shah's power and increase that of his office. His cabinet indeed included many figures who had reason to hate Reza Shah and the Pahlavi dynasty. As much to placate the Soviets as to gain the support of a tightly organized and powerful element, Qavam began by showing favour to the Party of the Masses, to the extent of giving them two cabinet posts; he also gave one portfolio to a representative of the equally leftist Democratic Party of Azerbaijan. Qavam also formed a left-leaning Democrat Party in anticipation of elections for the Assembly due at the end of 1946. He also set up a Supreme Economic Council with an ambitious programme for distributing crown lands to peasants, setting minimum wages, and protecting local industries. Being in control of the administration he would, of course, be well placed to rig the elections and produce an Assembly ready to do his bidding.

Qavam found, however, that control of the whole country was not within his reach. In October, tribal disturbances erupted in the south, and the army, controlled by the Shah, was antagonistic and unwilling to be used to quell disturbances. The tribes captured two southern towns, massacring their garrisons, while a body of 15,000 warriors marched on the city of Shiraz. The tribes demanded autonomy similar to that enjoyed by Azerbaijan, the exclusion of Tudeh from the government and a ban on its activities in the south. Qavam found that he had to distance himself from Tudeh and its leftist sympathizers, and he tried to make peace with the tribes and the landowners and businessmen and include them in his Party.

In the election, Qavam was able to rig the vote in favour of his Party in the capital, and in northern and eastern provinces where he had connections and influence. In the event, the Democrats won eighty seats, the court connection was able to muster thirty-five, and twenty-five members returned from southern constituencies where British influence was extensive, constituted a pro-British bloc. These parties and blocs were by no means monolithic, none less so than Qavam's supporters. The eighty Democratic members were a particularly disparate group, comprising both leftists and those whom Qavam had had to conciliate following the October disturbance. It did not take long for the Party to break apart. In the months following the convening of the Assembly in June 1947, members began to defect, and by October the cast-iron majority with which Qavam began became quite brittle. In a vote of confidence, only forty-five members supported the Prime Minister, thirty-six voted against, and thirty-one abstained. Qavam resigned and left for Paris, and the supporters of the Shah introduced a bill for his impeachment on the grounds that he had sold import licences in order to finance his Party. Qavam's second attempt at trying conclusions with the young Shah had failed.

A year or so afterwards, thanks to an accident, the Shah seemed set wholly to regain the power which his father had wielded. In February 1949, an attempt was made on his life. The would-be assassin was killed on the spot, but it was ascertained he was a journalist who wrote for a religious paper, and that he belonged to a Tudeh-affiliated union. This enabled the Shah to declare martial law, to ban newspapers critical of the court, and to outlaw Tudeh. He also convened a Constitutional Assembly which unanimously voted to establish a senate, half of whose members would be the Shah's nominees. It also gave him the right to dissolve Parliament at will, provided new elections were simultaneously ordered and the new Parliament would meet within three months. The royal estates which Reza Shah had accumulated and which were transferred to the state in 1941 now reverted to the Shah. A supporter of the Shah became Minister of the Interior and thus in a position to rig the forthcoming elections.

The elections, however, could not be properly rigged. Another opponent of the dynasty unexpectedly found in them his opportunity, and almost succeeded in bringing down the Shah. This was Muhammad Mosaddeq (1880–1967). Mosaddeq came from an aristocratic family and by 1951, when he became Prime Minister and began his challenge to the Shah, he had had a long administrative and political career, in this respect recalling Zaghlul, who only late in life made his distinctive mark on Egyptian politics. In 1921 he had served as Governor-General of the province of Fars, and when Reza Khan rose to the top between 1921 and 1925, he was one of his supporters, serving as Minister of Finance and of Foreign Affairs. He opposed, however, Reza Khan's elevation to the throne, and retired from public life. Like so many opponents of the regime, Mosaddeq came back on to the political scene in the years following Reza Shah's downfall. He was elected to the Assembly which convened in February 1944, and quickly established himself as a people's tribune—an opponent of the Shah's dictatorship, and a jealous defender of national independence. Western-educated, he also believed that Iran's future was linked with reform and modernization.

Like Zaghlul, Mosaddeq refused to be the spokesman of a party, claiming to represent the whole nation. However, during the elections which took place in the autumn of 1949 he, together with a crowd of supporters, led a march to the palace gardens to protest against the lack of free elections. The Shah was preparing to visit the US to seek aid, and it may have seemed to him that a liberal image would help him in his quest. Mosaddeq and his followers were promised that there would be no more vote-rigging. Following this success, a National Front was formed of which Mosaddeq became the leader. The founders of the Front were mainly lawyers, professionals, and journalists who, like Mosaddeq, were generally Western-educated and who believed in constitutionalism and in the possibility of transforming Iran into a modern and progressive society—all of these desirable goals having been hitherto obstructed by the Shah's despotism and the reactionaries on whom he leant. Among the groups constituting the National

Front there was one, however, which was quite different. This was the Society of Muslim Warriors which was led by a cleric, Ayatollah Abul-Qasim Kashani, who had a long past of political activism stretching back to the anti-British revolt of 1920 in the Middle Euphrates. The Society stood for the repeal of secularist laws imposed by Reza Shah, the reinstatement of the *shari'a* as the public law of Iran, the reimposition of the veil, and pro-Islamic solidarity. Associated with them was a small terrorist group, the Fida'iyan-i Islam, i.e. those who lay down their lives for Islam, who were to play a crucial role in Mosaddeq's ascent to power.

The issue on which Mosaddeq fastened in the Assembly was one which had come up in the negotiations beween Qavam and the Soviets in 1946—the issue of an oil concession for the Soviets in northern Iran. This in turn brought to the fore the issue of the Anglo-Iranian Oil Company which operated in the south, and with whom Reza Shah had negotiated a concession in 1933. The government now negotiated a new concession which it brought before the Assembly for approval in June 1950. The majority in the Assembly were supporters of the Shah, and the National Front had only eight members, including Mosaddeq. He and his supporters ferociously attacked the new concession as exploitative, and demanded nationalization of the oil industry. Such was the outcry against the oil company, orchestrated by Mosaddeq and his supporters inside and outside the Assembly, that the government felt unable to bring the matter to a vote, and instead resigned.

The Shah then appointed as Prime Minister General Ali Razmara, the Army Chief of Staff, in the expectation that he would be forceful enough to push through the new concession. Mosaddeq, of course, vehemently attacked Razmara before large crowds in Tehran, and Kashani called upon all sincere Muslims and patriotic citizens to fight against Razmara and to join the struggle for nationalization. In March 1951, Razmara was assassinated in the central mosque in Tehran by a member of the Fida'iyan. Shortly afterwards, Mosaddeq introduced a bill nationalizing the oil industry which the majority in both the Assembly and the Senate proceeded, in their panic, to approve.

The panic increased when the now outlawed Tudeh led a general strike in Khuzistan—the province where the Anglo-Iranian Oil Company operated—which was accompanied by sympathy strikes in Tehran and other cities. The upshot was that Mosaddeq was offered the premiership by a spokesman of the majority in May 1951. He accepted.

Mosaddeq, like Zaghlul, appealed to the mob. In order to intimidate the majority who, notwithstanding their vote in favour of his premiership, were in reality utterly opposed to him, and to carry through the nationalization, the Prime Minister encouraged public meetings and processions to which he appealed over the heads of the Shah and the Assembly. He had his way over nationalization, forcing the company to close down its installations and leave the country.

Elections for the Parliament were due at the end of 1951. Mosaddeq tried in his fashion to rig the elections by assigning literate and illiterate voters to different constituencies, and increasing urban representation, but his bill failed to pass the Assembly. When the elections were taking place Mosaddeq realized that they would result in a majority against him. He, therefore, stopped the elections midway, after 79 deputies, enough to constitute a quorum, had been elected!

It was this truncated Assembly which was to witness Mosaddeq's attempt to defeat the Shah and possibly bring to an end the Pahlavi dynasty and even the monarchy itself. The Assembly contained thirty of Mosaddeq's followers and supporters. It first met in February 1952 and the struggle between the Prime Minister and the Shah came to a head in July. Mosaddeq claimed the right himself to nominate the war minister, the Shah refused to accept his candidate, and Mosaddeq resigned on 16 July. The Assembly, where he did not have a majority, elected Qavam to succeed him. The National Front, however, aided by Tudeh, organized widespread strikes and demonstrations. The Shah called on the army to suppress them, but after five days of rioting and disorder in which twenty-nine rioters were killed, the Shah called off the army on 21 July and asked Mosaddeq once more to form a government.

Mosaddeq had triumphed. He declared the events which

brought him back to power to be a national uprising, and those who had fallen in the riots to be national martyrs. He transferred the Shah's lands to the state, appointed himself acting war minister, appointed one of his supporters court minister, and forbade any direct communication between the Shah and foreign representatives. In less than a year Mosaddeq had succeeded in reducing the Shah to little more than a figurehead. Mosaddeq also cowed the Parliament, by obtaining emergency powers for a period of six months which allowed him to legislate by decree on a very wide range of subjects. Members of the Senate were supposed to serve for six years: Mosaddeq forced through the Assembly a bill reducing their tenure to two years. When he found opposition in the Assembly inconvenient, he had his followers resign *en masse*, so that there was no longer a quorum, and the Assembly was automatically dissolved. In July 1953, the Prime Minister called for a popular referendum to approve his actions, on the score that the people of Iran created the Constitution and had full right to make any change they wanted. As might be expected, the referendum, in the August following, went overwhelmingly in Mosaddeq's favour, those approving his policies forming a classical 99.9 per cent of the voters. The exercise was distinguished by a novel and ingenious feature: there were separate ballot boxes, for 'yes' and 'no' votes, and the boxes for the 'yes' votes were in a different location from those of the 'no' votes. Mosaddeq, however, did not enjoy for long the authority conferred on him by popular suffrage. On 19 August, a conspiracy in which were involved army officers and British and US agents, successfully organized a popular tumult while a general surrounded Mosaddeq's residence with tanks and arrested him and his supporters. The Shah who, a few days before, had fled the country following the failure of another conspiracy against Mosaddeq shortly before, now, in his turn, came back in triumph. He then ruled almost unchallenged for a quarter of a century or so.

After his arrest Mosaddeq was tried for treason and sentenced to three years in prison. On his release, he spent the rest of his life under house arrest on his estate in Ahmadabad near Tehran. Mosaddeq's attempt to gain and retain power required the subversion of constitutionalism which he had always claimed to champion. In this respect, he and the Pahlavi dynasty were two similar sides of the same coin. As for his attempt to obtain greater prosperity for Iran through taking over the oil industry, this too proved a fiasco. Nationalization deprived Iran of the oil revenues on which it relied. Foreign exchange became scarce while inflation, shortages, and unemployment increased. In spite of all this, Mosaddeq continued to have many admirers among the intellectual classes who, during the long years of Muhammad Reza's absolutism, felt strongly nostalgic for Mosaddeq and the National Front, in the belief that if only they had succeeded an era of freedom and prosperity would have dawned for Iran. These illusions persisted until the Shah's downfall, both among the Iranian intelligentsia and among many influential figures in the United States, which had become the Shah's patron. Equally persistent was the belief that the National Front represented, in some sense, the wishes and aspirations of the Iranians. As events showed, during and after the Shah's downfall in 1979, this belief was equally false. The well-known poet Saless who was a supporter of Mosaddeq and of the National Front, dedicated to him a poem which, with its elegiac tone, captures well the intellectuals' regret and nostalgia:

> Did you see, my heart, that the beloved did not come?
> The stallion arrived; but the rider did not come?[24]

With Mosaddeq's downfall, Parliament fell under the Shah's total control. His powers were enlarged by his being given a veto over financial bills; elections were to take place every four, rather than two, years as had been the case from 1906 onwards, and the size of the quorum required to enact laws was lowered. Two parties were formed, the National Party (replaced in 1963 by the New Iran Party) and the Party of the People, presumably on the principle, first enunciated by the Khedive Ismail in the 1860s in Egypt when he ordered that the members of the assembly which he had set up should be divided into a government and an opposition party; otherwise it would not be as advanced and up to date as the European versions on which it was modelled. In a book published in 1961, the Shah declared that if he had been a dictator, he might have been tempted to 'sponsor' a single party; he was, however, he said, a constitutional monarch who could 'afford' to encourage 'large-scale party activity'. In 1975, the Shah changed his mind and abolished the two parties. He replaced them with a single party, the Resurgence Party:

The Resurgence Party was designed by two groups of very divergent advisers. One group was formed of young political scientists with Ph.D.s from American universities. Versed in the works of Samuel Huntington, the distinguished political scientist at Harvard, these fresh returnees argued that the only way to achieve political stability in developing countries is to establish a disciplined government party. Such a party, they claimed, would become an organic link between the state and society, would enable the former to mobilize the latter, and thus would eliminate the dangers posed by disruptive social elements . . . The second group of advisers was formed of ex-communists from Shiraz who had left the Tudeh in the early 1950s—one had absconded with the party funds . . . This group argued that only a Leninist-style organization could mobilize the masses, break down traditional barriers, and lead the way to a fully modern society.[25]

Only one untoward incident marred the precision of the electoral parade in the years following Mosaddeq's turbulence. In 1960 the Prime Minister and leader of the National Party agreed with his colleague, the leader of the People's Party, on a division of seats in the Assembly with the greater share going to the Nationals. So it came to pass. However, the vote-rigging was judged too blatant by the Shah. He advised the elected deputies to resign. The Prime Minister also resigned, and new elections were decreed for January 1961. By then, J. F. Kennedy had been elected President of the US and it was deemed politic to make a gesture to establish the Shah's liberal credentials: one National Front leader and some independent candidates were allowed to be elected.

With the passage of the years the control by the Shah and his officials over most aspects of the economy and of society went on increasing. A graphic indication of this centralized control is the vast increase in the official bureaucracy and in the concomitant increase in the number of desk-bound employees, which was mostly in response to the necessities imposed by the progressively encroaching centralization. The category of professional, managerial, and clerical workers increased from 272,000 in 1956 to 2,006,000 in 1976, while urban wage earners in commercial and industrial establishments increased between the same years from 1,441,000 only to 2,674,000. This intervention—and interference—in almost every aspect of social and economic life went on apace. As in Turkey, during the 1960s and 1970s this gave the Government the illusion that it could plan everything in great detail and that everything would come out as planned. It also, however, burdened the Government with responsibility for everything that could, and did, go wrong. Planning in the late 1950s led to deficits in the public accounts, to attempts to cover the deficits by printing money, to inflation, price rises, recession, discontent, and a mounting chorus of complaints against the Government.

The difficulties gave the Kennedy administration enough leverage over the Shah to persuade him, in exchange for financial relief, to embark on a land-distribution programme and to appoint in April 1961 as Prime Minister a former ambassador to Washington, Ali Amini, who had established a reputation for progressiveness. Amini's cabinet included radicals and former supporters of Tudeh and the National Front. The administration embarked on land reforms and adopted the stringent financial policies demanded by the International Monetary Fund in exchange for its help. The land reform, later adopted in a modified form by the Shah himself after he had dismissed Amini in June 1962, aroused the opposition of landowners, both lay and clerical. The stringent financial measures adopted at the insistence of the IMF led to increasing discontent by shopkeepers, wage-earners, and the unemployed. In particular, one cleric in the shrine city of Qum, Ayatollah Ruhallah Khomeini, became very outspoken in his attacks on the regime and on its subservience to the US and Israel. In March 1963, on the anniversary of the martyrdom of the Sixth Imam, Ja'far al-Sadiq, Khomeini's religious college in Qum was attacked by paratroopers and secret policemen. Students were killed and Khomeini arrested. He was released shortly afterwards and resumed his fiery attacks on the Shah and on the US as the enemy of Islam. On the anniversary of the martyrdom of Imam Husayn at the hands of the Umayyads, on 11 June, Khomeini was again arrested. The traditional processions held to commemorate Husayn's martyrdom turned into demonstrations and riots in Tehran, Qum, Shiraz, Mashhad, Isfahan, and Tabriz. The riots lasted for three days and left many hundreds dead. Khomeini was released in August, resumed his agitation, imprisoned again in October 1963, released again in May 1964, and finally exiled to Turkey, whence he went in 1965, to the shrine city of Najaf in Iraq.

During and after the October 1973 war between Israel and her Arab enemies, the Organization of Petroleum Exporting Countries (OPEC) suddenly decided to raise the price of oil five-fold. The Shah was one of the main ringleaders in this coup and Iran, a principal oil producer, increased its income from oil to prodigious heights. The new riches made the Shah embark on very ambitious programmes of industrialization, of expansion in the armed forces, on all kinds of social and economic planning. These policies on the one hand unleashed unprecedented inflation, and on the other attracted increasing numbers of migrants from the countryside to the city. They were attracted by the lure of high wages, but the cities to which they came had none of the municipal and social infrastructures necessary if the migrants were to lead at all tolerable lives in their new and alien surroundings. They came to form an ever-increasing discontented and volatile mass. When, in 1977, the usual effects of inflation and economic mismanagement began to show themselves, popular discontent became more and more manifest. As in 1961, there was now too a Democratic administration in Washington headed by President Jimmy Carter. One of their self-imposed missions was to make the Shah, in the eyes of many high officials in the new administration a sinister and oppressive despot, change his ways and become more liberal.

To please his patrons, the Shah again, as in 1961, decided to show a liberal countenance. He allowed public criticism of the regime, relaxed police control over dissident groups, amnestied a large number of political prisoners, and introduced new rules protecting the rights of defendants in political trials which, he promised, would be tried in civilian, not military courts.

These actions began to destabilize the regime. Its various political opponents emerged from decades of silence, and as the months went by and the regime seemed not to react in the usual repressive way, took courage and became increasingly outspoken. Khomeini, from his refuge in Najaf, stepped up his attacks which were circulated all over Iran in cassettes. The economic stringency which inflation had produced spurred attacks on the regime which became more and more vociferous. During 1978 as demonstrations turned to riots, the Shah seemed unable to follow a clear and sustained line of policy. Orders were given, then countermanded; at one moment the army was told to intervene, at the next to abstain. The army and the administration which he had made dependent on himself alone, became utterly disoriented. Apathy seemed to seize hold of him, whether because of the cancer from which he was suffering, or because he felt abandoned and betrayed by Carter and his officials. One of his last Prime Ministers had the brilliant idea of having Khomeini expelled from Najaf. The Ayatollah went to France where myriad journalists and television reporters magnified his message and in the end made him seem invincible and irresistible. As in 1953, the Shah decided to give in, and, as in 1953, he fled the country. On 16 January 1979, he took the plane to Cairo, never again to return. The fall of the Pahlavis also meant the end of the 1906 Constitution, and the promise it held of a Western-style constitutional and parliamentary government.

6

The Triumph of Ideological Politics

FOR a century or so, constitutional and parliamentary government was the form of political organization which the Middle East tried to adopt from the West. The rulers of the Ottoman Empire, of Egypt, Tunis, and Persia introduced constitutions and parliaments, not because the institutions were imposed on them by foreigners, but because the rulers themselves genuinely believed that they would be most beneficial.

If, after the First World War, it was the Mandatory Powers which set up constitutions and elected parliaments in Iraq and the Levant, this was not done against the wishes of the local governments or the official classes. On the contrary, after the Mandates came to an end, the now fully independent governments evinced no desire for change.

Constitutionalism was thus the Western political tradition which was adopted earliest in the Middle East. It had a long, albeit very checkered, career. It proved, as has been seen, to be a failure everywhere, with the possible and highly qualified exception of Turkey.

Constitutionalism was, however, by no means the only Western political idea to become familiar to the Middle East, and taken up by its political classes. Two kinds of Western ideas vied in popularity with constitutionalism, and were in the end to prove much more attractive, more consonant with certain strands of the native political tradition, and thus more significant in their impact. One such powerful, if for a long time submerged, Western idea is that of millennialism.

Belief in an eventual transfiguration of all history, when men shall see a new heaven and a new earth, is an early feature of Christianity. The promise of a new heaven and a new earth indeed appears in one of the books of the New Testament, the

Revelation of St John the Divine 21: 1. This millennialism was originally the belief that a thousand years after Jesus' first appearance, a Second Coming would take place, when love, liberty, and joy would reign on earth everlastingly. From the beginning, those Christian theologians who formulated the main elements of orthodox belief saw the dangers of millennialism and combated it. Long before the year 1000 (when this transfiguration was supposed to happen) St Augustine, in *The City of God*, dismissed this belief as 'fit for none but carnal men to believe'; the Council of Ephesus, also, in 431, condemned belief in the millennium. For Augustine, and for subsequent Catholic authorities, the Revelation of St John in no way tells us what the future has in store, but has to be read as an allegory of the soul's journey towards salvation.

Jewish and Muslim divines have also been very clear in condemning similar apocalyptic expectations. The reason, of course, is not far to seek. Such apocalyptic expectations must give rise to great social and political disorder, and will thus pose grave dangers to the security and the life of the faithful, as well as to their eternal salvation.

Condemnation by religious authorities notwithstanding, millennial expectations and apocalyptic movements could never be repressed. They led an underground and disreputable life in European society. They would now and again erupt in heresies, the proponents of which would seize on whatever discontents happened to be prevalent in society at large, would promise deliverance from present miseries, and sometimes attract a numerous and fervent following. By their help they would seek to subvert and overthrow the existing order, and endeavour, of course always in vain, to establish a new order in which universal salvation would obtain here on earth.

With the secularization of the European mind the millennial expectation articulated in religious terms came to be transformed into the utopian hope that men would enjoy gradual and continuous improvement in their earthly condition. But even when the idea of progress seemed to replace the millennial hope, the earlier vision could still find fervent expression. In a short work published in 1780, *The Education of the Human*

Race, which is a most significant document in the annals of European thought, G. E. Lessing (1729–81) proclaimed his belief that the course of human history was inexorably leading towards perfection. In his tract, Lessing showed that he was aware of the origin of his idea of human perfectibility in medieval writers who had been inspired by the Book of Revelation.

Nor is Lessing's vision a merely bookish fancy. The course of modern European history shows that powerful and far-reaching political movements had as a mainspring this tenacious belief that a new era of justice and happiness was about to be inaugurated. This was certainly the case with the American revolutionists who proclaimed a *novus ordo saeclorum*—a motto which figures on the Greal Seal of the United States and is reproduced on every dollar bill.

It was also widely prevalent in the French Revolution, as Wordsworth's well-known poem attests: for the young poet, the Revolution appeared as 'a pleasant exercise of hope and joy' when

> the whole earth
> the beauty wore of promise[1]

In the middle of the nineteenth century Giuseppe Mazzini entertained the same expectation from the inauguration of the era of nationalities. Again, in the twentieth century, the Bolshevik Revolution in 1917, the Fascist march on Rome in 1922, and the Nazi triumph in 1933, were thought by the votaries of each movement to inaugurate a new epoch.

As has been said, Christian millennial ideas have their analogues in Islam, specifically in the idea of a *mahdi*, that is a well-guided one, who comes at the end of time, when anarchy reigns everywhere, to banish injustice and institute equity. In Sunni Islam the idea did not become prominent. In addition, Sunni rulers had great reason to fear the spread of mahdist notions, promoted as they were by a variety of Shiites who challenged the legitimacy of these rulers, and claimed that the descendants of Ali, the Prophet's cousin and son-in-law, and in their eyes his sole legitimate successor, were the only rightful rulers of the Muslims. Sunni divines insisted that a *mahdi* cannot appear unless no Caliph reigned, and until the world was engulfed in disorder and arbitrariness. An authoritative legal treatise of the sixteenth century declared that the Imam (the legitimate Muslim ruler) 'must be visible; he must neither hide from the eyes of the public, nor be the object of its [messianic] expectation'.

The stipulation that the Imam must be visible is one clearly directed against Shiite tenets. The Twelver Shiites believe, as has been previously said, that legitimacy inheres in those descendants of the Imam Ali designated for the imamate by their predecessors, and the sole legitimate Imam in the world today is the twelfth descendant of Ali (hence their appellation as Twelvers)—a descendant who disappeared from mortal view in the ninth century, who is still alive, but invisible, but who will appear in his own good time to re-establish justice and equity in the world. Mahdism is thus a central feature of Shiism, but more often than not for the faithful the belief entailed not activism, but rather passivity. As Shiite divines taught them, believers should await this coming in patience and quietness.

Just as millennialism became secularized in the modern world, so did the idea of a *mahdi* become a tool for a secular sort of politics in the hands of one of the most influential Muslim thinkers in modern times. This was Jamal al-Din al-Afghani, whose role in the tobacco protest against Nasir al-Din Shah has been mentioned in Chapter 3. Notwithstanding the false name by which he wished to be known, and which deliberately gave the impression that he came from or was connected with Afghanistan, and thus probably a Sunni, Afghani was in reality a Persian and a Shiite. He would thus have been very familiar with the Shiite belief in the Hidden Imam, and in his awaited appearance. Afghani, however, changed significantly the original notion of the *mahdi*, moored as it had been in a long-hallowed religious tradition. This change may have resulted from exposure to Western ideas, with which he had most probably come in contact when he was in India and Central Asia in the 1850s and 1860s; or else, stimulated by these Western ideas, he may have worked it out for himself by drawing on certain heterodox currents within Shiism itself.

The transformation of the idea of the *mahdi* at Afghani's hands may be seen in a series of articles which he wrote for a French socialist newspaper in December 1883 following the Sudanese Mahdi's victory over a British expedition sent against him. Afghani declared that Muslims have a sure belief that a *mahdi* must come, that every Muslim awaits a *mahdi* for whom he is ready to sacrifice his life and all his possessions. The Indian Muslims, in particular, 'in view of their infinite sufferings and the cruel torments they undergo under English domination, await him with the greatest impatience'. England, Afghani also declares, hopes in vain 'to stifle the voice of the *Mahdi*, the most awesome of all voices, since its power is even greater than the voice of the Holy War, which issues from all Muslim mouths'. Does England, he asks,

think herself able to stifle this voice before making itself heard in all the East from Mount Himalaya to Dawlaghir, from north to south, speaking to the Muslims of Afghanistan, of Sindh and of India, proudly proclaiming the coming of the Saviour whom every son of Islam awaits with such impatience? *El-Mahdi, El-Mahdi, El-Mahdi!*[2]

All Muslims, Afghani concluded in the last of his three articles, 'await the *mahdi* and consider his coming as an absolute necessity'.

It is clear that Afghani's *mahdi* is not the superhuman being of Shiite eschatology, but a political saviour whose credentials consist not in descent from the Imam Ali, but rather in the ability to inspire the masses to political action, the success of which will indeed prove him to be the *mahdi*. Afghani thus secularized the idea, and he seems also to have understood what its political uses might be—at any rate, its uses in the kind of politics made familiar in Europe by the French Revolution, and which he himself attempted, somewhat unsuccessfully, to practise. Afghani was clearly attuned to modern mass politics and to the conditions which make it possible. In the last of his articles in *L'Intransigeant*, Afghani remarks that

Man is by nature given to exaggerating all news which comes to him from afar so that the figure *one* travelling from one mouth to another and augmented by public rumour soon ends up by becoming the figure *thousand*; the hillock comes to be considered a mountain. This is why at the announcement of the coming of a *mahdi*, the hearts of all those who are waiting for their liberation will be filled with great expectations and will overflow with joy and hope.

Rumour, with its powerful effects which Afghani so graphically describes, does not have in the modern world to depend only on word of mouth. The editor of *L'Intransigeant*, Henri Rochefort, recounts a conversation which shows how aware Afghani was of literacy as a political weapon, and of the new role which the newspaper press can play in mobilizing the mass for political purposes:

He used to say to me with his Asiatic subtlety, 'England had thought it a great act of policy to impose the English language on the Hindus, whether Muslims or idolaters, but she has made a tremendous mistake. Today they understand the newspapers which their conquerors publish and realize perfectly well the state of subjection to which they have been reduced.'[3]

Afghani also told Rochefort that he would be sending pamphlets and newspaper extracts 'by the bale' to the most obscure townships in order to develop in them 'the spirit of insurrection'.

This, on Afghani's part, was boastfulness, since there is no evidence of bales of pamphlets and newspapers being sent by him to the East. As a man of action, in fact, Afghani was on the whole a failure. He tried, unsuccessfully, to engineer the murder of Khedive Ismail and to raise an agitation against his son and successor, Tawfiq, who expelled him from Egypt. He had earlier tried to play a role in Afghan politics, probably as a Russian agent, but had to leave the country in obscure circumstances. Similarly, his activities in Istanbul had led to his departure for Egypt. In Persia, his agitation against Nasir al-Din Shah led to his deportation. At one point, he went to Russia, where he hoped to organize an anti-British rebellion in India, which came to nothing. And in spite of his fulminations against oriental rulers and their despotism, contained in a short-lived Arabic periodical which he published in Paris, he ended his life as a pensioner of Sultan Abd al-Hamid, used to promote the Sultan among his fellow Shiites in the shrine cities of Mesopotamia. His only clear success lay, as previously mentioned, in procuring—at a distance—the murder of Nasir al-Din Shah by a simple-minded follower who believed, interestingly enough, that Afghani—who, when he acknowledged his Persian origin, claimed to be a *sayyid*, i.e. a descendant of Imam Ali—was himself the *Mahdi*. True to his own conception of a *mahdi*, Afghani believed that violence was the only way to realize his purposes. A Persian friend of his recounted that he happened to visit Afghani at his house in Istanbul where he found him pacing his room, seemingly oblivious of his surroundings and shouting in a frenzy, 'There is no deliverance except in killing, there is no safety except in killing'.

Afghani's career holds for us many points of interest. In the same way as happened in Europe, a traditional religious eschatology is transformed into a secular one. This secular eschatological vision is accompanied, as is the case with modern ideological politics in Europe, by a frenetic activism lured on by the prospect of power gained through indoctrinating and mobilizing the masses. While these characteristics are certainly present in others, who later played a part in politics, Afghani's own writings do articulate with clarity this change in the world view of Middle Eastern political figures, and his activism is a harbinger of the contemporary style of Middle Eastern politics.

Afghani's secular frame of mind is worth considering further. There is little doubt that in religion he was a sceptic, not to say an unbeliever. While in Paris, he had a public exchange with Ernest Renan, the eminent historian of the origins of Christianity, who had been educated at a Catholic seminary and had later lost his faith. In a lecture at the Sorbonne in March 1883, Renan discussed Islam and science, and denounced Islam for stifling all free thought with a dogma imposed through terror, and perpetuated through hypocrisy. The lecture drew a response from Afghani in the *Journal des Débats*. Afghani avowed here his belief that religion was, on the whole, now a force for evil. All religions, he declared, are intolerant. It is true that Islam did seek to stifle freedom of thought, but so likewise did Christianity and all other religions. Some four years afterwards, describing his conversations with Afghani, which had in fact decided him to lecture on the subject, Renan called Afghani an 'enlightened Asiatic' and went on:

The freedom of his thought, his noble and loyal character, gave me the impression, while talking to him, that I had in front of me one of my ancient acquaintances, such as Avicenna or Averroes, or some other one of those great unbelievers who, for five centuries, upheld the tradition of the human spirit.[4]

That the 'enlightened Asiatic' had put behind him any concern with the ideals of the religious life as articulated, say, in the Prophet's teaching, is clear from other reports of his opinions. A Persian friend reported him as declaring:

There are two kinds of philosophy in the world. One of them is to the effect that there is nothing in the world which is ours, so we must remain content with a rag and a mouthful of food. The other is to the effect that everything in the world is beautiful and desirable, that it does and ought to belong to us. It is the second which should be our ideal, to be adopted as our motto. As for the first, it is worthless, and we must pay no attention to it.[5]

Again, one among many of his grandiose projects may have been a plan confided to his most faithful followers for the subversion of Islam. In a letter of 1883 his young disciple, Muhammad Abduh, quotes the master's 'sound rule' to the effect that the head of religion is to be cut only with the sword of religion. What this maxim signifies is that, in Afghani's estimation, religion—or rather the appearance of it—is necessary in order to destroy genuine belief in religion among the masses. If Afghani dreamt of a practical application of his esoteric doctrine, then his dreams clearly remained mere dreams. The destruction of religion is, however, not the only use to which a false show of piety may be put. Afghani saw that religion could also be made to serve yet another purpose—one which related to his view of the *mahdi* and his political significance. Since, as he could well see, the masses were attached to their religion, use could be made of this attachment to reach those political ends which stood high in his own estimation. Hence, in his exoteric discourse, particularly in the eighteen numbers of the periodical which he published in Paris together with Muhammad Abduh, he insisted on the vital necessity of Islamic union. For its title the periodical adopted a phrase used by the Prophet to describe the power of religious faith. The editors called it *al-'Urwa el-wuthqa* (*The Indissoluble Link*), the better to emphasize its general theme. Union was the only way to withstand the European assault which had already caused such ravages. Solidarity is the foundation of power, hence the Koran had exhorted men to be united in clinging to His power and protection. Where Muslims were devout and punctilious in their religious duties, the Muslim state was strong, because religion ensured their unity and created their solidarity. It was the solidarity produced by Islam which made the disparate and disunited tribes into a force formidable enough to conquer a vast empire. It is with this train of ideas that Afghani is generally associated, and the reason why so many—both Muslim and non-Muslim, including some eminent orientalists—have been deluded into regarding him as a great Islamic figure.

For Afghani, then, Islam is above all a solidarity-producing force, and in this its value primarily lies. It cements society, makes it into a concrete-like phalanx resistant to all assaults, and itself able to conquer and overwhelm weaker, less compacted groupings. If solidarity is the supreme value, Afghani shows himself to have an open mind whether Islam, or some other belief, is more efficient in producing and maintaining it. In conversation with a friend during his last years at Istanbul, he is led to consider the comparative value of nationality and religion in this respect. Men, he said, are naturally found in groups, which are in conflict with one another over security and the control of scarce resources. This conflict forces these groups to unite according to the bonds of kinship, and form nations such as the English or the Russians. This is national solidarity. In Islam, a revealed religion provided a solidarity higher than the various national solidarities which had bound together the groups which came to adopt Islam. In an article written in Persian, Afghani, however, seemed to think that there is nothing stronger than national solidarity. He prefaced this article with an epigraph to the effect that there was no happiness except through nationality, and no nationality except through language, and in it he argued that a national unity based on a common language was more powerful and more durable than one based on a common religion.

2. The Lure of Political Violence

Afghani may stand as an exemplar, uniting in his own person the restless search for the secret of political power in the modern world and the frenzied, unscrupulous activism which this quest inspired. Many others among the official and educated classes, affected by contact with the West, exhibit the same beliefs and inclinations; but in few, if any, can this particular amalgam be so well-documented and apprehended so clearly. The Young Turk officers who carried out two *coups d'état* against Sultan Abd al-Hamid in 1908–9 exemplify many of the characteristics which are combined in Afghani. These officers and their seniors, stationed in Salonika, could see with their own eyes the violence which then raged in the Macedonian cockpit, pitting Greek against Bulgar and Macedonian against both. One of the most fearful terrorist organizations in the modern world, the Internal Macedonian Revolutionary Organization (IMRO), originated in the ruthless ideological struggle of which Macedonia was the scene. We catch a glimpse of this unscrupulous and violent style of politics in the obituary of an IMRO leader—perhaps the last surviving one—Ivan Mihailoff, who died recently at the age of 94:

In the 1920s and early 1930s, when he waged terrorist attacks on Bulgarian Government forces from a mountain hideout, Mr Mihailoff was known as the 'bandit king' of Macedonia. His Internal Macedonian Revolutionary Organization, or IMRO, was said to have killed 3,500 of its enemies in a twelve-year struggle that included factional fighting among rival Macedonian revolutionaries.

Among the killings was the 1925 assassination in Vienna of Tudor

Panizza, a fellow Macedonian leader, by Mr Mihailoff's wife, Menicha Karnitcheva. Mr Mihailoff himself acknowledged that he had ordered the execution in 1928 of Alexander Protogueroff, another Macedonian chieftain who had been held responsible for the assassination of Tudor Alexandroff, the top leader of the revolutionaries, four years earlier.

The Mihailoffs contended that many of the killings for which their organization was blamed had in fact been carried out by Serbians and Bulgarians seeking to suppress Macedonia's independence movement.[6]

The Ottoman army in Macedonia was not simply an onlooker; it was involved in maintaining, so far as it could, the authority of the Ottoman administration in the province, and to do so it had to find an answer—often an answer in kind—to terrorist and guerilla activity. So that the young officers who were to carry out the *coups d'état* had become familiar with, and inured to, violence which went beyond the limits of legality—whether the law in question was the civil or the military code. Their actions when in power after 1909—the assassinations of opponents, the conspiracies, the rigged elections, the ready recourse to intimidation through extreme and frightful acts, as when one of the Young Turk leaders, Enver (who thereafter became the dominant figure in the Ottoman government), burst into a cabinet meeting with an armed band, one of whom shot dead the minister of war—may be seen as the continuation, on a different stage, of the habits and methods which they had learned in Macedonia. The brutalization of politics in the Young Turk era is evident in many ways, big and small. To take a small but significant instance, a great many Young Turk-affiliated newspapers began to be published after the *coup* of April 1909, when Abd al-Hamid was deposed, with titles such as Weapon, Bayonet, Bullet, Knife, and Bomb.

Enver's record exemplifies the adventurist character of ideological politics. The ambition to make a reality of grandiose ideals can lead to violent and destructive action, far removed, in its cynical lack of scruple and its brutality, from the fervent belief that these ideals will of themselves guarantee a happy and shining future. Between 1913 and the end of the World War in 1918, the Young Turk leaders, of whom Enver was the most forceful, controlled the destinies of the Empire, and led it into a disastrous war. When it had to surrender to the Allies, the leaders fled abroad. Enver went to Moscow. He began by collaborating with the Bolsheviks in Central Asia. He then turned against them, and led a Muslim anti-Bolshevik movement, moved by the vision of a Turkic Power in Central Asia headed by him, which might, who knows, sweep away Russian domination, and even perhaps effect a junction with their Turkish brethren in Anatolia, thus redeeming and making a reality of his original ideals. Enver found his end in 1922 when he was killed in a clash between Soviet forces and the group he led.

The lure of the Macedonian example for the Young Turks is captured for us in one chapter of an opuscule written by a Young Turk ideologue which is significantly entitled 'The Idealists'. Those he holds up for his readers' admiration are doctors, lawyers, professors, 'and similar idealists' who have sacrificed their professional careers for the sake of a political cause:

First let us take the case of a young Bulgarian. He came of a wealthy family and was therefore pampered and indulged in every possible way by his parents. Up to the age of thirty he studied at the University. Does he then move to a luxuriously appointed lawyer's office in Salonika, surrounded by all the most modern of comforts? No! His office is in the mountains, his desk is a rock, his pen a gun and a dagger, and his clients Turkish gendarmes and Greek robber bands upon whom he passes sentence of death without much ceremony.

Another illustration. The young man has studied medicine in a European University, and returns home with his doctor's degree. Will he now declare war on the microbes and the thousands of diseases which assail human life? No, indeed. He wanders, armed to the teeth, from village to village, from mountain to mountain, dispensing out his only medicine, those death-dealing blue pills, to all the opponents of his ideal, and even to those of his countrymen who do not share his ideas.

A third picture is afforded by a professor of the highest philosophical attainments. Does he establish a centre of training in Athens, Bucharest, Sofia, or Belgrade? No again! In secluded villages such as Grebena and Dikvesh, etc., he instils the Irrendenta principles into the minds of the village children, and prepares them to sacrifice life and fortune for this ideal.[7]

The 'idealists' were so admired by Tekin Alp because they were pursuing, at all costs to themselves and to others, the realization of an ideal—a political ideal. Politics seen as the pursuit of ideals is ideological politics, which sits very uneasily, if at all, with what the government of an actual society requires. Ideological politics in pursuit of the ideal, on the contrary, demands, if necessary, the total destruction of an existing political order. It is this uncompromising attitude which was shared with their Macedonian adversaries by the Young Turk officers. It is an attitude which the success of the Young Turk *coups d'état* made very attractive to their fellow-officers in the Ottoman army, for instance the officers hailing from the Arabic-speaking provinces. Such an attitude is incompatible with a constitutionalist kind of politics. The failure of constitutionalism in the Ottoman Empire when government was in the hands of these officers is thus not surprising. Mustafa Kemal too, himself a Young Turk by conviction—though not a prominent political figure during the decade of Young Turk dominance—had his own ideal, or blueprint, which circumstances and his own genius enabled him to impose on Turkish society. Here too, the absence of constitutionalism in the Turkish Republic during his reign and many years after his death, is therefore not surprising.

3. Nationalism and Revolution

Mustafa Kemal had the sagacity and the good fortune to establish and maintain a stable regime immune from *coups d'état* and similar upsets. The partiality to violence and conspiracy characteristic of the Young Turk officers, however, infected their fellow officers who came to power in Iraq after the First World War, and the same attitude was transmitted by example to younger officers. Such attitudes became even more pronounced among them in successive decades, as the corruption and failure of the regime became increasingly manifest. Military *coups d'état* in Iraq from 1936 onwards constituted a vicious circle. The leaders of every *coup* promised to cleanse the state from the corruption of its predecessors; the promise naturally could not be fulfilled; and rivals, making the same promises, in their turn toppled those in power. The destruction of the monarchy in 1958 did not spell an end to this turbulence: it continued for a decade, until a military *coup d'état* established a regime much more ruthless than its predecessors in suppressing incipient challengers.

The same radical state of mind was manifest among army officers in both Syria and Egypt. Conditions in these two countries were somewhat different from those obtaining in Iraq. In both countries, it was civilian notables who were at the forefront of political activity, against the French mandate in Syria, and the British occupation in Egypt. Here too, however, the rulers who inherited from the French and the British respectively showed themselves both incompetent and corrupt. Owing to their proximity to Iraq and to the closer connections between the two countries, the Syrian officers were much more exposed to the radical ideologies which moved Iraqi officers. But even though the educated and official classes of Egypt were comparatively unaffected by the intellectual climate which reigned in the Levant and Iraq, and though their preoccupations were somewhat different, yet it did not take long for young Egyptian officers to be moved by impatience with the existing regime, and by a belief that if only power were in their hands, the lot of their fellow-Egyptians would be improved out of recognition.

Iraqi ex-Ottoman officers, as has been said, acquired from the Young Turk experience what might be called a culture of conspiracy, lawlessness, and violence. These tendencies were greatly reinforced from the 1930s onwards through the spread of European radical ideologies, which came to be greatly and widely admired. This was most true of Nazism in the 1930s. The toughness it preached, the quasi-military discipline it sought to instil in its followers, the ideological indoctrination it systematically pursued, its vision of politics as violent conflict in which winners naturally would 'liquidate' the losers, the total renovation of society to which it aspired—all of this seemed to young officers, as well as to the teachers through whose hands they passed as schoolboys, to be supremely worthy of emulation. The founders of the Ba'th Party in Syria—all of them secondary school teachers during the 1930s—did not hide their admiration for the ideology of German nationalism as preached, for instance, by Fichte, or for its modern Nazi version. Also in the 1930s, an Iraqi Director-General of Education, Dr Sami Shawkat, addressing secondary-school boys exalted what he called the 'Manufacture of Death':

The nation which does not excel in the Manufacture of Death with iron and fire will be forced to die under the hoofs of the horses and under the boots of a foreign soldiery. If to live is just, then killing in self-defence is also just. Had Mustafa Kemal not had, for his revolution in Anatolia, 40,000 officers trained in the Manufacture of Death, we would not have seen Turkey restoring in the twentieth century the glories of Yavuz Sultan Selim [known in European historiography as Selim the Grim]. Had not Pahlavi had thousands of officers well-versed in the sacred Manufacture we would not have seen him restoring the glory of Darius. And had Mussolini not had tens of thousands of Black Shirts well versed in the Manufacture of Death he would not have been able to put on the temples of Victor Emmanuel the crown of the first Caesars of Rome . . .

The spirit of Harun al-Rashid and the spirit of al-Ma'mun [caliphs who reigned when the Abbasid empire was at its apogee] want Iraq to have in a short while half a million soldiers and hundreds of airplanes. Is there in Iraq a coward who will not answer the call? . . . If we do not want death under the hoofs of the horses and the boots of the foreign armies, it is our duty to perfect the Manufacture of Death, the profession of the army, the sacred profession.[8]

Sami Shawkat's colleague, Dr Fadil al-Jamali, Inspector General of Education, deplored that Iraq had nothing to compare with the Hitler Youth or the Communist *Komsomols*, organizations which, in his belief, demanded of the young 'faith, discipline, and united action'—the fundamental conditions of political success.

Nazism, the shining example of the 1930s, did not last as long as Mustafa Kemal's or the Pahlavis' regime. After its downfall, however, other radical ideologies swept Asia and Africa after their European rulers, one after another, had departed. If Jamali was rather exceptional, during the 1930s, in his admiration for the *Komsomols*, two decades later the Soviets had replaced the Nazis as a source of emulation. They too had an alluring doctrine about rebuilding or renovating the whole of a society, about banishing forever injustice and inequality within, and between, countries, and doing it through revolution.

After the First World War there had come into existence all over the Middle East, small, clandestine, and generally insignificant Communist Parties, composed mostly of intellectuals, which proved desperately ineffective in trying to organize workers or fellahin, and lead them to revolution. It was obvious that here was no industrial proletariat which could serve as the vanguard of the revolution, according to the Marxist schema. Some officers in Egypt, Iraq, and Syria did belong to the Party, but it was not this which made the Soviet example significant. What, rather, the Soviet Union was believed to show, in all its might and world-wide influence after 1945, was that revolution could indeed turn a society round, and make it efficient and powerful. In the Middle East, not the proletariat, but the officers were the real vanguard. This is the alluring dream which moved Nasser and which finds expression in *The Philosophy of the Revolution*. That Nasser seemed, for a while, to be able to enact the dream made him a model and a hero to countless officers in Iraq, in Syria, in Libya, and elsewhere in the Arab world.

The idealism of the Macedonian guerrillas which Tekin Alp so much admired was focused on a political doctrine: nationalism. The doctrine was invented in Europe at the turn of the eighteenth and nineteenth centuries, and its central idea was that humanity was naturally divided into separate and distinct nations, each identified by its own language, customs, and history. Each nation, the doctrine also held, had to live in its own sovereign state, as otherwise the identity of its members would dissolve and disappear in that of another nation—the greatest misfortune which could befall them. This doctrine, which looked for the happiness, indeed the salvation, of man to a political arrangement, spread out of Europe to the whole world, and wherever it spread, came to be embraced with a burning, fanatical fervour. 'Seek ye the political kingdom' as Nkrumah said, parodying the New Testament, 'and all the rest shall be added unto you.'

One of the earliest areas outside Europe to which the doctrine spread was the Ottoman Empire, where the Christian subjects of the Sultan were more aware of European intellectual currents, and more easily affected by them. The fact of subjection thus began to seem increasingly intolerable, when compared to the splendid prospects, held out by the doctrine, of a fulfilled and happy life in the bosom of one's own independent nation. Nationalism, then, was the second strand of the European political tradition profoundly to affect Middle-Eastern politics.

The spread of this doctrine was, however, bound to result in explosive situations, and in drastic political and social disruptions. To give reality to the ideal of nationalism meant, of necessity, challenging and engaging in violent struggles with an existing regime which stood in the way of its fulfilment. Also, very frequently, the attempt to gain independence according to nationalist criteria took place in areas where the population was extremely mixed, as for instance in Macedonia where various nationalist movements mercilessly fought one another, and where the Young Turk officers learned their earliest political lessons.

The Greek-speaking areas of the Balkans, for centuries under Ottoman domination, were the first to challenge this domination in the second and third decades of the nineteenth century. The Greek uprising, helped by sympathetic European Powers, led to an independent Greek state. It was followed by other uprisings in Ottoman territories in the Balkans which, owing to European support and Ottoman weakness, also led the establishment of independent states supposedly on the national principle. Also, towards the end of the century, an Armenian nationalist movement attempted an armed challenge against the Ottomans in Eastern Anatolia, where Armenians were numerous but not predominant, and where they lived in the midst of Turkish- and Kurdish-speaking Muslims. These challenges to the Ottoman dynasty and state, and to Muslim predominance in the Empire, could not but affect profoundly and in various ways the Ottoman official and educated class. In due course, increasing numbers among them came to consider that the bases on which the Empire rested, namely Islam and the Ottoman dynasty, were not adequate to create and maintain the social cohesion and the loyalty indispensable if the state were to be saved from dismemberment and catastrophe, if it was to stand any chance to become as powerful and prosperous as the states of Europe.

There had thus to be a transvaluation of values. Those who had hitherto seen themselves to be Muslims and Ottomans now came to think of themselves as Turks; descendants, that is, of those tribes of Central Asia who had occupied Anatolia, as well as other far-flung regions, and whose history and language would provide the authentic foundation of a renewed national existence. Ironically enough, this new enthusiasm for Turkish-ness and its virtues derived from the labour of Western Turcol-ogists whose philological and historical researches revealed the past achievements and abiding characteristics of the Turkish nation. How necessary this transvaluation was is made clearer by Zia Gökalp (1876–1924), one of the most influential expo-nents of Turkish nationalist doctrine. In articles written before and during the World War, he explains how the non-existence of the idea of nationalism among the Turks has led to economic and political disintegration:

As the non-existence of the ideal of nationalism among the Turks resulted in the lack of any national economy, so the same factor has been an obstacle to the development of a national language and to the appearance of national patterns in fine arts . . . The notions of solidarity, patriotism, and heroism did not transcend the confines of the family, the village, and the town.[9]

Gökalp goes further, and argues that the diversity of peoples co-existing in the Empire dooms it to have no future. In Europe, he explains, as a result of the introduction of universal military service and the suffrage, it became necessary to instil in the masses a sense of patriotism and teach them civic responsibility:

When the needs of adult and universal education became apparent, conflicts arose among the different ethnic groups in the state over the question of which language should be spoken in the schools. The government began to insist on the dissemination of an official language, but each ethnic group demanded that its own language become the main channel of education and instruction . . . Today in Europe only those states which are based on a single-language group are believed to have a future. Every national group is demonstrating the kind of future to which it aspires by voicing its wishes for a national home, with or without an historical basis.[10]

What, however, was the Turkish nation, which was now to be revivified as the focus of loyalty and solidarity? As has been said, the notion of Turkism owed a great deal to the researches of European scholars. Some of the scholars argued that the Turks belonged to a much larger ethnic group, including Mongols, Finns, and Hungarians, whom they called Turanian, after Turan, an ancient Iranian name for the area lying to the north-east of Persia. For Gökalp the Turks formed part of a much larger nation. In a poem of 1911 he declaimed:

> The country of the Turks is not Turkey, nor yet Turkestan,
> Their country is a vast ancestral land: Turan!

and in another poem written after the Ottoman entry into the war in 1914 Gökalp prophesied that 'Turkey shall be enlarged and become Turan'. This pan-Turanian dream it was which, a few years later, led Enver to his fatal Central Asian adventure.

The ideology of Turkism or Pan-Turanism was clearly incompatible with the Ottoman scheme, in which Islam and the dynasty were the foundations of the state, and which encompassed a large number of religious and ethnic groups. Furthermore, no one could tell what the boundaries of a Turkish or Turanian nation might be. In the event, the slaughter-bench of history took care of the issue. From the Greek revolt onwards, in the course of a century or so, after a great many wars, massacres, and the uprooting of long-settled populations, the Ottomans gradually lost all their possessions in the Balkans. In

Anatolia, however, they managed to defeat nationalist claims, made first by the Armenians and then by the Greeks. The Armenian challenge in Eastern Anatolia during the 1890s led to what came to be known as the Armenian Massacres, and later on, during the World War, to the deportation and the killing of almost all Armenians in Anatolia whom the Ottoman Government suspected of planning an uprising in support of a Russian invasion from the Caucasus. After the Ottoman defeat and surrender in 1918, the Greek Government, with Allied support, sent an expeditionary force in order to annex Smyrna and its hinterland where large numbers of Greeks were settled. In the course of the ensuing war, massacres and atrocities were committed on both sides. The invasion ended disastrously for the Greeks, who had to evacuate the Anatolian mainland. In the settlement made at the Conference of Lausanne 1922–3 an exchange of populations between Greece and the Turkish Republic was agreed, and between 1923 and 1930 a million and a quarter Greeks were sent from Turkey to Greece, and a somewhat smaller number of Turks from Greece to Turkey.

At first sight this exchange seems a clear indication of the prevalence on both sides of nationalistic and patriotic ideas, and the desire to give greater unity and cohesion to the nation and the fatherland. Yet on closer examination of what actually took place, it begins to appear that other ideas and other loyalties were still at work. The Greeks of Karaman who were 'repatriated' to Greece were Greek Christians by religion—yet most of them knew no Greek. Their language was Turkish—which they wrote in the Greek script . . . In the same way, many of the repatriated Turks from Greece knew little or no Turkish, but spoke Greek—and wrote it in the Turco-Arabic script . . . A Western observer, accustomed to a different system of social and national classification, might even conclude that this was no repatriation [of Greeks and Turks] at all, but two deportations into exile—of Christian Turks to Greece, and of Muslim Greeks to Turkey.[11]

That these deportees were now to be called respectively Greeks and Turks is an eloquent, and to those concerned, a painful, example of the transvaluation of values which nationalist doctrines required, and nationalist struggles imposed.

The disappearance of the Armenians and the ejection of the

Greeks from Anatolia made possible a colourable case for arguing that the unit over which Mustafa Kemal was in control following his victory over the Greeks was not simply that part of the old Ottoman domain which had not seceded or been conquered, but in reality was and had long been a country called Turkey. In opting for this reading of the past and the present, Mustafa Kemal rejected not only Islam and the dynasty as the foundation of the body politic. He also rejected the alternative nationalist vision of Turan as being the primordial unit of which the Turks of Anatolia were one component. In a speech, as early as December 1921, he dismissed Pan-Turanism as a fantasy which could not be realized, and urged a return to 'our natural, legitimate limits'.

Mustafa Kemal was lucky in that the country he now ruled and which was officially given the name Türkiye—a very recent coinage—in 1923 came, through accidents of war and civil war to be composed in large part of Turkish-speaking Muslims whom the government could use all its resources to persuade that in reality they were the Turks depicted in the official ideology. Nationalism became one of the six principles of Kemalism, incorporated in the constitution. According to this Kemalist doctrine, the Turks were Aryans from Central Asia, where all civilizations had originated. The Turks in due course had migrated to various parts, and brought the arts of civilization with them. They thus founded Chinese, Indian, and Middle-Eastern civilizations. In the Middle East, the Sumerians and the Hittites were in reality Turks, and Anatolia, where the Hittites founded civilization 4,000 years before the Christian era, was thus Turkish from prehistoric times. Turkish man, as Tekin Alp, who became an exponent of the new history, wrote, quoting a French authority,

is one of the most beautiful specimens of the white race, big in stature, with an elongated and oval face, a fine nose, either straight or aquiline, sensitive lips, eyes opening widely, quite often grey or blue, and with horizonal palpebral slits.[12]

However, even after the virtual disappearance of the Armenians and the Greeks, these beautiful specimens of the white race were not the only inhabitants of Turkey. In the east and south-east of the country lived substantial numbers of Kurds. Accurate estimates are difficult to secure, and while the authorities are intent on minimizing their numbers, the leaders of the Kurds are as naturally intent on maximizing them. To say that Kurds in Turkey number anywhere between five and ten million would give one an idea of the order of magnitude involved. For various reasons, religious, economic, and ethnic, Kurds have proved a restive element in the Turkish Republic. In the Ottoman Empire the Kurds were one element among many in a multi-ethnic and multi-religious polity. Islam was the dominant element in this polity, and as Muslims, Kurds felt at home in it. This was not the case in a secularist republic, where Islam was depreciated and where citizenship was deemed to be closely connected to, if not actually defined by, Turkishness. The authorities attempted to suppress all mention of Kurdish identity, and repressed rebellion in Kurdish areas which erupted in the 1920s and 1930s.

The desire to repress and suppress has indeed taken many forms over the decades. One recent ingenious attempt is the Language Prohibition Act of 1983. The 1982 Constitution had declared that Turkish was the language of the state, and had made provision for banning publications and speeches 'in languages forbidden by law'. The 1983 Act, presumably based on this constitutional provision, prohibited as a criminal offence the use of any language which is not the official language of a state recognized by Turkey in the terms of international law. Kurdish clearly fell under this ban.

The Kurds, for their part, were now touched by the idea of Kurdish nationalism. The idea of Turkish nationalism was inspired and evoked by nationalist movements appearing in the Empire. Kurds likewise follow the same nationalistic logic adopted by the governments under whose authority they find themselves. This logic means that unless they become Turks in a Turkish national state, they must be content with an inferior position. However, it is manifestly very difficult if not impossible for Kurds—or Greeks or Armenians—to look upon themselves, and to be accepted, as Turks. What is true of Turkey is *mutatis mutandis*, equally true of other states in the Middle East which are erected upon the nationalist principle; Kurds, again, in Iraq where Arabism is the foundation and aspiration of the state; or Arabs in Israel, the emanation of Zionism, an ideology where nationalism is predicated upon Jewishness.

Kurdish nationalism, like all similar movements, aspires to establish a Kurdish national state. The aspiration was fed and encouraged by the political discourse which became current during the First World War, and which became even more entrenched after 1945. In articulating their war aims, the Allied Powers, and President Woodrow Wilson in particular, used widely the idea of national self-determination as the basis on which a new, and better, world order would be built. The Kurds were among those peoples who were to enjoy the benefits of national self-determination. The treaty of Sèvres between the Allies and the Sultan, signed in August 1920, made provision for a Kurdish state. The Treaty, however, was never ratified by the Grand National Assembly of Ankara (which had assumed the prerogatives of the Sultan). The Treaty of Lausanne which replaced it did not provide for a Kurdish state, but the idea has never died among Kurdish leaders, whether in Turkey, Iran, or Iraq. In the case of Iraq, particularly, the League of Nations, in approving the ending of the Mandate and the establishment of a fully sovereign state, required that the Kurds should enjoy a degree of autonomy. Iraqi governments never met this requirement in any real sense, and relations between Baghdad and the Kurds have been uniformly tense, and often violent, under successive regimes. There were many tribal rebellions under the monarchy, as under the military regimes which followed it. A rebellion led by Sheikh Mustafa Barzani, and carried out with the help of the Shah of Iran, collapsed when the Shah suddenly abandoned the Kurds in 1975, following a deal with the Iraqis. The consequences for the Kurds were quite catastrophic, since many of them were forcibly moved from their areas and settled in the south of the country, under very difficult conditions. When the Iraq–Iran war started in 1980, the Iranians once more helped and incited the Kurds of Iraq to rise against Baghdad. This led to severe and continued repression by the Iraqi government, which went on with its policy of deportation, and also used poison gas against Kurdish civilian populations. The pursuit of a Kurdish national state led also, as has been seen, to the setting up during the Second World War, with Soviet support, of a short-lived Kurdish republic in Mahabad. It also led to disturbances and uprisings following the Shah's downfall and the war with Iraq, when Iraq, like Iran, attempted to foment troubles for its enemy in the Kurdish region.

The fate of the Kurds in the last few decades points up one feature of ideological, and particularly nationalist politics which, from the Armenian Massacres onwards, has become increasingly prominent in the Middle East. Political and military conflicts are no longer confined to governments and armies. The civilian, generally unpolitical, population becomes a target in the conflict—a target for the ideologists who wish to indoctrinate and mobilize, and thus equally a target for governments and groups who may feel threatened by the ideology in question. Hence a whole population may be either deported or massacred, to be replaced by one which is politically reliable. This was true not only of the Kurds, but also of the various Lebanese sects, which became one another's mutual victims in the civil war which began in 1975, and of the conflicting ideologies in terms of which the contending parties articulated their respective cases.

4. Arab Nationalism and Arab Socialism

In spite of their contrived and artificial character, Kemalism and Turkish nationalism were, on the whole, accepted by the bulk of the population for whom they were meant, and have not, so far, set up insoluble tensions in the society of the Turkish Republic. This, however, was not the case with Arab nationalism. Just as Turkish, or Pan-Turanian, nationalism was inspired by the spread of nationalist ideologies in the Ottoman Empire, so Arab nationalism, in its beginnings, came out of the Young Turk *coup d'état*, the political tensions which followed it, and the ideological style of politics which the Young Turk officers practised and made popular.

During the long years of his rule, Sultan Abd al-Hamid gave special attention to the Arabic-speaking provinces. Abd al-Hamid took great care to emphasize the Islamic character of his office. He was not only the Ottoman Sultan, but also the Muslim Caliph. This meant in turn emphasizing the importance of the Islamic Holy Places, Mecca and Medina, as well as the fact that the capitals of the most ancient Islamic empires, Damascus and Baghdad (now in reality backwaters), had long been part of the Ottoman realm. Abd al-Hamid also surrounded himself with important officials and religious figures from the Arabic-speaking provinces who were believed to wield much power and influence. When he was deposed in 1909, his policies and his counsellors shared in the discredit which his successors eagerly heaped upon his regime. Officers had been recruited in increasing numbers from the Arabic-speaking provinces, owing to Abd al-Hamid's policy of setting up military secondary schools in provincial centres. These officers had the same outlook and the same radical mind-set as their Turkish colleagues who formed the secret Committee of Union and Progress and who carried out the *coup d'état*. When it seemed as if these Young Turks nursed an anti-Arab animus, the Arab officers, no doubt carried away by the great effervescence of the times, started forming secret societies, in imitation of the Young Turks. Their aims, although not precisely articulated, combined hostility to Turkish or Turanian nationalism (which various publicists were now quietly promoting), with demands for autonomy or independence in the Arabic-speaking provinces.

Before the outbreak of war in 1914, these secret societies were able to accomplish nothing. Secession from the Ottoman Empire came to the fore through British attempts, in the course of the war, to foment trouble from the Ottomans in the Empire. This would weaken Ottoman war-making ability. In addition, if Ottoman domination in the Arabic-speaking provinces was weakened or destroyed, then this opened up prospects of an increase in British power and influence in the area. This was very much something which Lord Kitchener, the Secretary of State for War, desired. He had been the British representative in Cairo between 1910 and 1914, and during his tenure of this post, there had been approaches from Husayn, the Sherif of Mecca, who sought to obtain British support in order to strengthen his position *vis-à-vis* the Istanbul government. Following Ottoman entry into the war, Kitchener directed that the pre-war contacts should be taken up. On his instructions, Husayn was offered the prospect of becoming Caliph. As previously said, the Caliphate which Kitchener and the officials in Cairo had in mind was a so-called spiritual Caliphate, whereby Husayn would become a kind of Islamic Pope who sat in Mecca and who would somehow exercise religious authority over the Muslim world. An office of this kind was utterly unknown to Islam, where the Caliph was—to use Western terms—the temporal, territorial sovereign as well as the religious head of the Muslims. The negotiation begun at Kitchener's initiative was taken up, after a while, by Husayn. In letters exchanged between him and Sir Henry McMahon, Kitchener's successor in Cairo, during 1915–16, Husayn asked that Britain 'acknowledge' Arab independence over a very large area including the Levant, Palestine, Mesopotamia, and the Arabian Peninsula. The British did not—could not—agree to such a demand, and the correspondence continued in a desultory fashion, and came to no conclusion. Husayn, however, rose against the Ottomans in May 1916, took captive the Ottoman garrison in Mecca, and with the help of the British navy, established himself in Jedda and on the Hijaz coast. Husayn was financed, armed, and advised by the British, and supported by contingents of Muslim troops from Egypt and Algeria, and the Ottomans proved unable to extinguish his revolt. Arab officers, who had joined the secret societies formed in the years immediately preceding the war, deserted to the Sherif and served in his forces, which were in effect tribal levies. Faysal, the third son of the Sherif, whom Colonel Lawrence took up, and whose cause he tirelessly promoted during and after the war, became the commander of the so-called Northern Arab Army. This force was used by General Allenby to harass units of the Ottoman Fourth Army in TransJordan during the summer of 1918, while he himself faced and routed its main body in battles in northern Palestine and southern Syria. At the end of September, Allenby's troops were at the gates of Damascus. Allenby, as has been said, then forbade all his troops, except the Bedouins and the Druse who formed Faysal's small force, to enter the city. This was in order that Faysal should claim Damascus as his conquest. This, the British hoped, would deny the French control over Syria— control to which they had agreed in the Sykes–Picot Agreement signed in 1916, but which they now disliked.

For about a year and ten months, following his installation in Damascus, Faysal became the ruler of Syria. Faysal was able to do so, and to defy the French, who were claiming their rights under the 1916 Agreement, owing to the protection of British troops, who garrisoned Syria until towards the end of 1919. He also enjoyed British diplomatic support, which continued until the British were faced with having to choose between him and the French. In March 1920, a few months before the French expelled him, Faysal's followers defiantly proclaimed him King of Syria. Among the most influential of Faysal's followers, who were given important posts in the Arab administration, were ex-Ottoman officers and officials, including those who had deserted to his father during the war. They saw Syria as a base from which they would be able to work for the single Arab state of which the secret societies had dreamed before 1914, and which the Sherif had demanded in his correspondence with McMahon. During their brief period in power, they fomented disturbances in British-occupied Palestine, which they called Southern Syria.

Shortly after Faysal's expulsion from Syria and the destruction of his regime, as has been said, he was offered, and he accepted, the throne of a newly-created Kingdom of Iraq which the British had set up in Mesopotamia, in a desire for financial retrenchment, and under the persuasions of Colonel Lawrence. With Faysal there came to the new kingdom those ex-Ottoman officers and officials who had been the pillars of his regime in Damascus and who, here too, took up the reins of power. Like its short-lived precedessor in Damascus, the Kingdom of Iraq would become the base whence Arab nationalism would be spread over the Arab world, effecting a change in the traditional values of this world, and attempting to create a unified Arab state: as Arab nationalists repeatedly affirmed in the 1930s, Iraq was to be the Prussia or the Piedmont of an Arab world awaiting its Bismarck or its Cavour.

It was in Iraq that a fully-fledged doctrine of Arab nationalism was articulated. The resources of the state were employed to spread the doctrine among the young in schools and colleges, the curricula of which were officially prescribed, and where teachers were chosen who were imbued with the idea, and could be trusted to spread it among their charges with a fervent zeal. The most prominent author of the doctrine, and whom Faysal appointed as the first director-general of the Ministry of Education, was an ex-Ottoman official who had served him as Minister of Education in Damascus, and who came to Baghdad in his train. Sati' al-Husri (1879–1968), born in Aleppo, and trained as a pedagogue, had occupied various posts in the Ottoman educational service. Though he is said never to have lost his Turkish accent in speaking Arabic, he became the first ideologue of Arab nationalism, whose ideas and influence spread all over the Arab world. This was undoubtedly greatly helped by his prominence in Baghdad during the 1920s and 1930s, which gave his teaching added prestige and resonance.

Following the Rashid Ali affair in 1941, the British determined to clean up the Ministry of Education, which they considered to be a centre disseminating nationalist, anti-British and pro-Nazi sentiments. Husri was stripped of his Iraqi nationality and expelled from Iraq. But his influence by no means diminished. He became an education official in Syria, and was then appointed head of the cultural department of the Arab League. From the end of the Second World War until his death he published a large number of books which spread his ideas all over the Arab world. It is said that it was owing to Husri that Nasser became converted to Pan-Arabism and made it an ideological pillar of his regime—which it remained until his death in 1970.

Husri was concerned to rebut attacks by traditional divines for whom Islam was the supreme value, and who considered

that the affirmation of a national identity by Muslims constituted a danger to Muslim cohesion and solidarity. His argument
was that Arab unity had to precede Islamic unity: if Arab
Muslims were united in one state, this would actually facilitate
Muslim unity; to oppose Arab unity would be in fact to hinder
union among Muslims. Husri, however, went further, and
denied that Islam, or any religion, could supply the foundation
for a political structure. Husri, echoing the ideas of the German
originators of nationalism, affirmed that the individual finds
complete fulfilment only in merging himself with the nation and
in practising complete obedience to the state. In an address to
a nationalist club in Baghdad in the 1930s, Husri affirmed that

he who refuses to annihilate himself in the nation to which he belongs
may, in some cases, find himself annihilated within an alien nation
which may one day conquer his fatherland. This is why I say continuously and without hesitation: patriotism and nationalism before and
above all . . . even above and before freedom.[13]

That Arab nationalism is the supreme value to which everything and everyone in the Arab world should be subordinated
became a recurring motif in nationalist discourse. An Iraqi
ideologue of the 1930s, Muhammad Mahdi Kubba, described
by a contemporary Egyptian journalist as the Goebbels of the
nationalists, voiced, probably under Husri's influence, the same
aspiration to a total unity in which all Arabs would be unified
in feelings and opinions, and share the same hopes. This would
be done, Kubba argued, by planting the same principles in the
Arabs' souls. The regimentation and indoctrination which, as
has been seen, Sami Shawkat and Fadil al-Jamali preached at
that time was directed to the same end, namely to make the
individual wholly absorbed in the nation. Another, the Syrian
Ali Nasir al-Din, who had also been in Baghdad in the thirties,
wrote in 1946 that Arabism is the religion of the Arab nationalist, and that the individual Arab should feel that he is the
nation and that the nation is himself. The best known nationalist thinker following Husri, Michel Aflaq (1910–89), who
began his career as a schoolteacher, expressed the same view—
a view which he had worked out for himself under the influence

not of Husri, but of the German thought which had inspired Husri. Aflaq demanded a complete re-fashioning of Arab society, which would transfigure the life of the Arab. This requires the cultivation of hatred—hatred and annihilation of everyone whose ideas stand in the way of this transfiguration. This hatred, however, is in reality love—love for Aflaq's fellow Arabs who must be brought to see that, deep down, and unbeknown to themselves, they harbour the same aims as the nationalists, even though their actions may betoken the contrary. Nationalists must practise a loving mercilessness towards their brethren; nationalism is indeed, he wrote, 'love before everything else'. Nationalists, he said,

are merciless to themselves, merciless to others. If they discover in their own views a mistake, they correct it without fear or shame . . . If they find the truth anywhere, then [for its sake] the son will deny his father and the friend abandon his friend.[14]

The same celebration of violence is found in the writings of Frantz Fanon, the ideologue of the Algerian *Front de Libération Nationale*. Fanon was born in the French West Indies and trained in France as a psychiatrist. Working in a hospital in Algeria during the anti-French rebellion which erupted in 1954, he became convinced that the mental disorders sent for treatment to his hospital were the result of colonial oppression. He left his post and joined the rebellion of which he became a well-known spokesman. In a long essay, 'Concerning Violence', written in the late 1950s (he died in 1961 at 37 years of age), Fanon lyrically praised violence as a cleansing force which bestows freedom on colonized man. Violence, Fanon declared, is a cement, 'mixed with blood and anger', which binds the society of the oppressed struggling for their freedom: it 'binds them together as a whole, since each individual forms a violent link in the great chain, a part of the great organism of violence'. The practice of violence will end by creating a homogeneous and brotherly society. In Fanon's case, as in that of all the above writers cited here, the European influence is patent. In one way or another, these ideas all derive from the millennialism which, as has been seen, has deep roots in European thought. In its modern garb, adopted by figures like Fanon or Aflaq, it issues in the belief that salvation is to be found in the life of politics and that revolution is the instrument by which it is attained.

Aflaq, together with his fellow schoolmaster, Salah al-Din al-Bitar, founded the Ba'th, i.e. Resurrection or Renaissance, Party in Damascus in 1940–1. The Constitution of the Party, published about a decade later, affirmed its belief that freedom of speech and assembly, of belief and of artistic freedom are sacred. It declared that sovereignty is the property of the people, which is the source of all authority. The regime of the united Arab state for which the Party was working was to be constitutional and parliamentary and it was to ensure the independence and total immunity of the judiciary. The Party, the Constitution also laid down, was revolutionary, and its aims could not be achieved

except by means of revolution and struggle. To rely on slow evolution and to be satisfied with a partial and superficial reform is to threaten these aims and to conduce to their failure and their loss. This is why the party decides in favour of:

1) The struggle against foreign colonialism, in order to liberate the Arab fatherland completely and finally.

2) The struggle to gather all the Arabs in a single independent state.

3) The overthrow of the present faulty structure, an overthrow which will include all the sectors of intellectual, economic, social, and political life. (Article 6)

The founders of the Party apparently never considered that the revolutionary enterprise to which they committed themselves, and which entailed a complete overthrow of all social and political institutions, was not compatible with freedom of speech and assembly or a constitutional and parliamentary regime. There can, however, be no doubt that it was the lure of revolution, and its promise of the violent destruction of the unsatisfactory and confining present, which attracted Aflaq and Bitar, and drew their schoolboy followers who were the original nucleus of the Ba'th. Revolution and revolutionary action was the promise of all those who challenged existing regimes in the Middle East. Nasser expounded his ideas and aims in a pamphlet which was, inevitably, entitled *The Philosophy of the Revolution*, while Khomeini's movement, so different in aims and assumptions from both Ba'thism and Nasserism, advertised itself as the Islamic Revolution.

A revolution which creates a new Arab life requires that its goals should be defined, and its energies guided towards those goals. This is the function of the leader, who has a vision of the aim to be attained, and who will guide the people towards it. A leader like Nasser, or a 'vanguard party' like the Ba'th, required total control of all the institutions of a society. For them, constitutional shackles, or a society the members of which were free to pursue each his own interests and satisfy their material or spiritual desires, would be an encumbrance and a hindrance. For such a leader or such a party the people is at best a plastic and malleable material to be fashioned into the desired, ideal form. An Arab nationalist, certainly not as well known as either Husri or Aflaq, the Lebanese Abdullah al-'Alayili, writing in 1941, emphasized the necessity of a 'powerful and violent' leader if Arab unity was to come about. Such a leader, self-confident, charismatic, ascetic in his habits, and spiritually united with his people, will be able to arouse the mass—which is otherwise occupied with frivolous things, like a child—by appealing to their emotions and then controlling and directing them to the desired goal. 'Alayili's words uncannily foreshadow the appearance of Nasser and similar leaders, as well as their techniques of domination, down even to the obligatory claim of leading a simple, modest, ascetic life: Nasser in his modest suburban villa, or the Iraqi Abd al-Karim Qasim snatching a few hours of sleep on a portable bed in his office, so as to devote himself day and night to the welfare of his people. Again, Nasser's oratory, and the electronic media which greatly magnified its power and immeasurably increased the size of its audience, is a graphic illustration of 'Alayili's point that such a leader has to arouse the mass and lead it towards the aim on which he himself has fixed. A follower of the Tunisian leader, Habib Bourguiba, describing the effect of his oratory on the masses, speaks—quite accurately—of a union in which the soul of the leader is 'mixed' with that of the masses, to the point, even, that when he compels them to think, they will use their minds. Politics here ceases to be a debate between leaders and their fellow-citizens, or even the impersonal transmission of orders as in a hierarchy. It becomes, rather, what Aflaq has described and foreshadowed when, developing the notion that nationalism is love, he declared, in reference to politics that

any action that does not call forth in us living emotions and does not make us feel the spasm of love, the revulsion of hate, that does not make our blood race in our veins and our pulse beat faster is a sterile action.[15]

Constitutional government was bankrupt in Egypt, Syria, and Iraq. It also proved bankrupt, after a much shorter experience, in Libya and the Sudan, where parliamentary governments had been inaugurated under Western guidance in 1951 and 1956 respectively. In Libya, a military *coup d'état* put an end in 1969 to the monarchy, and 1958 saw the first of many *coups d'état* in the Republic of the Sudan. From the 1950s onwards, therefore, the officers had the destinies of all these countries in their hands. The regimes which they established were even more centralized than their predecessors'. Having issued from *coups d'état*, the new regimes necessarily felt insecure and hence tried to keep all the reins of power very tightly in their own hands. The new rulers also believed that they could quickly ameliorate the material conditions of the people by establishing a command economy, which would break up large landholdings for the benefit of the landless peasants, and which would quickly obtain the benefits of industrialization through central planning. The professed aim was to establish socialism, in the belief that socialism was the most productive and equitable social system. The language used in the Constitution of the Ba'th Party expresses the ideals which the officers generally entertained. The Party, Article 4 declares,

is a socialist party. It believes that socialism is a necessity which emanates from the depth of Arab nationalism itself. Socialism constitutes, in fact, the ideal social order which will allow the Arab people to realize its possibilities and to enable its genius to flourish, and which will ensure for the nation constant progress in its material and moral output. It makes possible a trustful brotherhood among its members.

The officers also firmly believed that the purity of their motives and the intensity of their fervour would absolutely ensure that the Arabs were delivered from foreign domination, and corruption and oppression at home, and attain full unity.

Following a *coup d'état*, then, in each country the officers would inaugurate a new agricultural and industrial order in which grandiose plans would be announced to create prosperity and welfare for everyone. The plans and the new order they heralded came under the rubric of socialism, or Arab socialism. A representative expression of these aspirations occurs in the Egyptian constitution promulgated by Nasser in January 1956, which declared that the state was committed to economic planning and social welfare, and in the National Charter six years later, which set up a single party—the only legal one—called the Arab Socialist Union. Also, in 1964, a provisional constitution based on this Charter was more explicit than the earlier one. It declared that the system of government was 'democratic socialist . . . based on the alliance of the working forces of the people', and that the economic foundation of the state was the socialist system.

A first grandiose plan launched immediately after the *coup d'état* of July 1952 was a land-reclamation project in what was called the Liberation Province in the Western Desert. After a great deal of expenditure, the project came to naught. Government land-reclamation projects elsewhere were also generally very disappointing. Immediately upon seizing power, the military regime also embarked on 'land reform', whereby large holdings of land were broken up and redistributed to small farmers and landless peasants. However, this compulsory transfer of property benefited only a small proportion of the rural population. Breaking the landowners' power has redounded not to the benefit of the small peasant, but to that of the centralized state which has thus removed one further obstacle to its omnipotence. It is the bureaucracy, as Galal Amin has remarked in his strikingly entitled book, *The Modernization of*

Poverty (1974), which has tended to inherit the power of the landlord.

Further, agricultural output was impeded by a veritable swamp of price regulations and subsidies linked to a succession of five- and ten-year plans, supposed to effect the industrialization of Egypt. The net result of these operations was by no means to increase the welfare of the beneficiaries of land reform. It led, rather, to a substantial drop in agricultural production. Between 1970 and 1980, dependence on food imports increased greatly: for example, for wheat from 44 to 72 per cent, for sugar from zero to 43 per cent, for beef from 3 to 33 per cent. Since the economy was unable to produce enough export goods to pay for these imports, the deficit was made up by grants, chiefly from the United States. As the years went by, the population became accustomed to low food prices, made possible by the extensive subsidies which the official planners had introduced. It proved quite impossible to remove or even decrease these subsidies. When, under pressure from foreign lenders, an attempt to do so was made, disturbances ensued, as happened notably in the bloody riots of 1977. Failure in agriculture was accompanied by failure in industry—notwithstanding the successive impressive-sounding plans. This failure became particularly evident when, after the Suez affair of 1956, the regime embarked on successive nationalizations of large industrial and commercial enterprises. To set the policies, and to manage the enterprises, new official administrative structures were created: the Permanent Council for the Development of National Production was followed by the Economic Organization, which was followed by the High Planning Council and the Committee of National Planning. These organizations fearlessly assumed central responsibility for the investment decisions which, in an unplanned economy, are the outcome of market conditions and entrepreneurial judgement about the likely profits and risks of each venture. The result of this politicization was to smother the nationalized industries in a web of suffocating bureaucratic controls with their hallmark of dilatoriness, inefficiency, and corruption. P. J. Vatikiotis, describing conditions in Egypt as Nasser left them after his death in 1970, speaks of the 'deplorable' condition of public administration, with its 'overmanned and wasteful bureaucracy' which Nasser's successor, Anwar al-Sadat, had to contend with in trying to unclog the economy and encourage foreign investment:

It is not only a matter of non-existent telephones, virtually absent telecommunications, inadequate water and power services which hinder the development of a free market economy or economic development, in any area from agriculture to industry. It is also the quality of the human factor in the equation. Ever since Nasser—for political reasons—opened up higher education to all and embarked upon an ambitious educational programme for which the country had neither the financial nor [the] human resources, the State has had each year to find jobs for thousands of university graduates, who normally expect to be absorbed into the state bureaucracy or state-owned enterprises. Security of tenure regardless of performance is not the best incentive to productive work. It is this obstructionist monster of the state machine with its red-tape, lethargy, and choking legalism which has impeded a faster rate of foreign capital flow into the country . . . it is literally a tremendous chore to exist in, say, Cairo on a daily basis. No individual or corporate investor who expects some return on his investment will put up with these haphazard and frustrating conditions.[16]

Syria and Iraq under military rule have, *mutatis mutandis*, followed the same agricultural and industrial policies as Egypt under Nasser. They have made successive attempts, following various *coups d'état* and in response to factional quarrels, to distribute land to peasants, and have similarly nationalized industrial and commercial enterprises. The economic results of 'socialism' have been likewise a disaster. Land reform in Syria, with the attendant bureaucratic controls, actually led in the period 1962–71 to the area of cultivated land declining, and likewise the yield of the principal crops, cotton and cereals. Swampy land in the hinterland of Hama, the Ghab, again, was drained and reclaimed, and it became fertile. It was distributed to small farmers. However

In 1969 a large-scale peasant uprising occurred in the Ghab over the question of repaying back debts to the Agricultural Bank. Peasant indebtedness stemmed largely from the state's sytem of exploiting the land. The size of the plot rented by the government to the peasant was too small to enable him to make a profit. The government rented the land to the peasant by the year and periodically redistributed it. This system transformed the peasants into agricultural labourers and prevented improvements in the land and cultivation.[17]

The uprising in the Ghab did not appreciably change the state of affairs in Syrian agriculture. Writing in 1980, the late Michel Seurat (subsequently murdered while a so-called hostage in the Lebanon) pointed out that the Syrian state, imposing its monopoly over all agricultural produce, itself arbitrarily fixed the price at which it would buy from the agriculturalist. Thus, for example, while the official price of a kilo of lentils was 125 piastres, the peasant did not receive more than 80 piastres, on the pretext alleged by the buying official that the lentils were not of good quality. The bureaucrat's arbitrary powers led to his illicit enrichment. The official in charge of the economy within the Syrian Ba'th was dubbed by the public as Mr Five Percent. In 1977, in order to counter persistent accusations of illicit gains, President Hafiz al-Asad and other officials who emulated his example generously made a gift of all their possessions to the state. They and the state, however, as Seurat pointed out, were one.[18]

What was true of socialist agriculture was even more true of socialist industry and commerce in Syria. Eighty to ninety per cent of large enterprises were nationalized in 1965 and later as a result of internecine struggles among Ba'thist officers. Nationalization politicized these monopolistic enterprises and smothered them in bureaucratic regulations. The enterprises were overstaffed and their productivity was very poor. They were in effect being milked by the bureaucracy. Out of 251,000 employees in the state sector, only 70,300 workers were productive. The price at which their production was sold mulcted the consumer, who, in the absence of competition, had willy-nilly to buy what the nationalized industries offered. The unit cost of a locally produced pencil was four piastres and it sold at twenty-five. Socialism in effect meant the dominance of the Ba'th, and since the Ba'th in Syria came, after 1966, to be equated with the Alawite sect, socialism, Seurat declared, meant the dominance of the Alawites over the Sunnis, in economic as in other matters.

Land reform, nationalization, and socialism also came to Iraq after the military *coup d'état* of 1958. This was followed by other *coups* in 1963 and 1968, interspersed with and followed by plots, intrigues, and quarrels among power-holders and between them and those who aspired to supplant them. Some of these *coups* and upsets led to sweeping decrees, meant to inaugurate a better era and eliminate injustice and corruption. The results were more or less similar to what obtained in Egypt and Syria. Land distribution and detailed bureacratic regulation of farming and agricultural prices were imposed. Iraq, a country with much fertile soil, was before 1958 largely self-supporting in food. Today, with the agricultural labour force double what it was in 1958, Iraq has to import 70 per cent of its food. Unlike Egypt and Syria, however, Iraq disposes of large revenues from oil which, owing to the great increase in oil prices brought about by OPEC after 1973, showed a vast increase. The income from oil, which went entirely to the government, was used predominantly to create a powerful military machine, and to embark on grandiose projects which were of very doubtful economic value. The government, thus, writes Patrick Clawson, ordered detailed plans to be prepared for

a Baghdad subway system, Mosul airport, 3,000 km. of railways, six-lane expressways to Turkey and Jordan, two large dams . . . an 1800 MW power station . . . an oil refinery . . . a 1–2 million ton iron and steel complex, a 200,000 ton aluminium smelter, a factory to make one million tyres per year, plus the Petrochemical Plant No. 2 which has gained notoriety for its chemical war capabilities.

These plans were far beyond the country's financial capacity unless phased over twenty years . . . Unable to set priorities, Saddam [Husayn, President of the Republic] had insisted on pressing ahead on every front simultaneously, much as he demanded developing simultaneously a broad range of sophisticated weapons. So many resources were wasted that many projects were left incomplete. More importantly, so many resources were channelled into large-scale projects that few resources were left for the small-scale investment much more vital to growth.[19]

So it is that, blessed with so many advantages, Iraq has yet remained a poor country, burdened with policies which waste its resources.

5. Egypt, Syria, Iraq: The Style of Military Rule

The officers who swept away with such blithe confidence the existing regimes found it by no means easy to govern with legality and in tranquillity, just as their record in economic management proved dismal. In Egypt the Free Officers—as they called themselves—were soon involved, shortly after the *coup d'état* of July 1952, in a struggle for power among themselves. The *coup d'état* was carried out in the name of General Muhammad Naguib. The King was sent away, the political parties were abolished, as well as the Constitution of 1923. To replace it by another constitutional instrument was in the event judged too risky, and the officers decided to rule through a Revolutionary Command Council. The abolition of the old order had, however, unleashed forces which threatened military rule, the most dangerous of which were the Muslim Brethren, who had a large following. The officers, again, proved to be riven by ferocious rivalries. The real authors of the *coup d'état* were a group of youthful officers, headed by Colonel Nasser. They had regarded Naguib simply as a figure-head. The figure-head, it turned out, had ambitions of his own and meant himself to rule. Some of the officers, for reasons of their own, took up his cause against Nasser's faction. Disturbances in Cairo ensued, and it nearly came to an armed clash between the two factions. From these manœuvres Nasser emerged triumphant, and sixteen officers from the losing side were tried before a Revolutionary Tribunal. Nasser, now Prime Minister, placed the officers who took his side at the head of most of the ministries. In October 1954, there was an attempt on Nasser's life by a member of the Muslim Brethren, which was said to be part of an extensive conspiracy. A People's Tribunal, composed of three of the Free Officers, was set up which tried a large number of the Brethren, sentenced six of them to death, and ten to long prison sentences. Naguib was alleged to be part of the conspiracy, dismissed from his office as President of the Republic and put under house arrest. Nasser now ruled alone and his companions in the *coup d'état* now became his clients and dependants. He was elected President of the Republic in June 1956. To mobilize the population against the Muslim Brethren in 1953, he had set up a Liberation Rally. This now became the National Union, to be transformed in 1962 into the Arab Socialist Union. These organizations were in no sense genuine political parties representing various interests in the country. They were rather appendages of the regime, recognized and treated by the population as such. The defence and protection of the regime was entrusted to a secret police apparatus, which with the passing years, and with tuition by foreign experts, became ever more complex, labyrinthine and tentacular. It is curious and interesting that the first lesson in controlling the population through a secret police was given by an American political scientist in the employ of the Department of State. In a paper on 'Power Problems of the Revolutionary Government' the political scientist taught his officer-pupils that the police

should be 'politicized' and should become, to whatever extent necessary, a partisan paramilitary arm of the revolutionary government.

He also taught that

The nerve centre of the whole security system of a revolutionary state (or of any state) lies in a secret body, the identity and very existence of which can be safely known only to the head of the revolutionary government and to the fewest possible number of other key leaders . . . It is the duty of this body to be aware of all prejudicial activity, incipient or actual, whether it involves cabinet ministers or a captain in the armed forces.[20]

This kind of secret-police apparatus, on the model of a Gestapo or a KGB, has become the hallmark of Middle Eastern military regimes.

Following his nationalization of the Suez Canal in July 1956 and the Suez war in which the United States procured for him a political triumph over Great Britain, France, and Israel,

Nasser extended his field of operations. By means of radio broadcasts beamed to the Arab world, his regime tried to elicit popular enthusiasm among the masses, and to destabilize the governments of Arab states which stood in his way. His object now was to create a Pan-Arab state of which he would become the undisputed leader. In 1958 it looked as if this ambition was near realization. In January 1958, Ba'thist officers in Damascus, fearing that internal and external plots might put an end to their dominance, decided to ask Nasser to establish a union between Syria and Egypt. Nasser gave his assent provided that the union was a 'total union', i.e. that he would become the undisputed master of Syria. The officers agreed and union was proclaimed on 1 February. The names of Egypt and Syria disappeared. In the United Arab Republic, they were known as the Southern and the Nothern Region respectively. Nasser visited Syria where he received an enthusiastic and tumultuous welcome and, standing by Saladin's tomb in Damascus, vowed to follow his example and 'realize total Arab unity'.

The union was based on incompatible expectations. Nasser clearly thought that the Syrian officers and political leaders would defer to him and be his subordinates, and that he would be able to rule Syria as he was ruling Egypt, and that no opposition to or even criticism of his policies would be allowed. As he would not share power in Egypt, it was out of the question for him to do so in Syria. The Ba'thists, regarding themselves as the guardians of Pan-Arab pieties, and believing that Nasser was beholden to them for giving him Syria on a plate, and thus making him the most powerful figure in the Arab world, did not see matters this way. Criticism and opposition were not slow in coming, both from the civilian leaders of the Ba'th, and much more significantly, from the officers. Nasser and the Ba'th became estranged from one another. In 1959, Nasser appointed his fellow-conspirator and relative by marriage, Major Abd al-Hakim Amir, now a Field-Marshal, as his viceroy in Damascus and tried to impose on Syria the kind of police control which obtained in Egypt. Syria became a kind of Egyptian colony. In 1963, when it was once again independent and Ba'thists were again in government, there was an attempt to revive the past and to create an even more extensive Arab union, now to include not only Egypt and Syria, but Iraq as well. A Syrian delegation visited Egypt to conduct talks on this scheme. In one of the sessions, a Syrian Minister tried to explain to Nasser what had gone wrong after 1958. A secret police, the Minister said, was of course necessary for the safety of the state, in order to prevent and undo foreign plots.

The anti-revolutionary saboteurs in the Arab countries are many and they must be watched and suppressed. But to depend solely on the Investigation Bureau is dangerous. What is the Investigation Bureau? It is a reserve force of the organized popular force. But there was no organized popular force, and therefore the Bureau was the only one in action.[21]

The 'Investigation Bureau' proved powerless to prevent a conspiracy and a military *coup d'état* which ended Egyptian domination in September 1961. Nasser, however, insisted that Egypt should still be called the United Arab Republic—as it remained known until his death in 1970. His successor, Sadat, restored the country's ancient and historic name, not, however, quite abandoning the new identity which had signalized Nasser's ambition: the country's new name was the Arab Republic of Egypt.

One year after his eviction from Syria, Nasser seized another opportunity to fashion a large and powerful Arab state. A *coup d'état* in Yemen in September 1962 replaced the monarchy with a republic. The new regime was, however, unable to establish its authority in a country riven by tribal and sectarian dissensions, and where the royalist cause was supported by Saudi Arabia with money and arms. The republican regime appealed to Nasser, who sent an expeditionary force in its support. His calculation must have been that if he were established in Yemen, he would be able to use it as a base from which to subvert and eventually control Saudi Arabia and its oil riches. Egyptian troops remained in Yemen for five years fighting, in difficult mountainous terrain, an elusive enemy whom not even the use of poison gas could subdue. After great expenditure of men and treasure, Nasser was forced to evacuate the Yemen following his defeat at the hands of Israel in the Six Day War of June 1967.

The Egyptian regime was not overthrown following this catastrophic defeat (and neither was that of Jordan or Syria, who were the allies of Egypt) and Nasser himself remained at the head of affairs, bowing to the overwhelming public demand (as evidenced by large well-organized demonstrations) that he should continue in office. Nasser then sacrificed, no doubt regretfully, some of his most prominent followers, including the Minister of War, to the public ire. They were tried and sent to prison. His relative, Amir, who was Commander-in-Chief, engaged in a conspiracy against him. He died, it is said, by his own hand.

Nasser himself died in 1970. To judge by the public expressions of grief in Egypt and in the Arab world, his loss was deeply felt. His populist rhetoric, spread everywhere by the electronic media, the defiance and intransigence which marked his style of rule, the luck which attended his spectacular strokes of policy: the Anglo-Egyptian treaty of 1954 which ensured British evacuation of the Canal Zone, the arms deal with the Soviet Union in 1955, the nationalization of the Suez Canal in 1956 and the discomfiture of the British and French which followed their attempt to topple him, the proclamation of the United Arab Republic, his patronage of Palestinian resistance to Israel, the building of the Aswan Dam with Soviet help, all this created for him an aura which no other leader has enjoyed before or since. It is, however, by no means clear that the furious activism which marked his regime in the end accomplished much—if anything—for the welfare of the Egyptians or for the international standing of Egypt.

Nasser was succeeded by Anwar al-Sadat, one of the conspirators of 1952. Sadat had also been involved in plots during and after the Second World War. During the war, he was a member of a pro-Axis group of Army officers who sought to establish contact with Rommel in Libya. Two German spies caught in Cairo confessed that he was among those who had tried to help them. He was arrested and kept in prison, but eventually released because the prosecution could not produce evidence sufficient to condemn him. He himself, however, has confessed in his autobiography that he was indeed involved with these spies. He was also arrested in 1946 on suspicion of complicity in the murder of an Egyptian personality accused of being pro-British. Again, according to his autobiography, the suspicion was correct, but when he was finally put on trial in 1948, the evidence was insufficient to convict him. After 1952, during most of Nasser's ascendancy, he was not part of the ruling inner circle. When Nasser died, he was Speaker of the National Assembly, a sinecure with no power and little influence. Upon Nasser's demise, his power collectively devolved, for the time being, on the members of his court. Each one of these figures, controlling a security or party fief, had his own ambitions which conflicted with those of his rivals. This may have been the reason why they chose Sadat for the succession. He had no fief of his own, and was thus thought unlikely to pose a threat to any of them.

Within a year, the barons were proved to be conclusively wrong. Quickly and methodically Sadat dismantled what he called the 'power centres' and sent those who controlled them to jail. Sadat now reigned supreme. For all the noise they had made in their heyday Nasser and Nasserism were now relegated to obscurity. Sadat did away with some of the bureaucracy which was the concomitant of 'socialism'. He undid some of Nasser's confiscatory legislation, and returned their property to the victims. He also gradually dismantled the Arab Socialist Union, and finally abolished it in 1980. In its place he established the National Democratic Party, which became the government party with the lion's share of seats in the People's Assembly, which a new Constitution, promulgated in 1971, had established. This Constitution made provision for other parties to be formed, and Sadat allowed a Socialist Labour Party, and a Socialist Liberal Party to exist. In the 1979 elections, the National Democratic Party obtained 320 seats, the Socialist Labour Party, 20, 13 of whom defected to the National Democrats or sat as Independents, while the Socialist Liberal Party obtained only two seats. Other parties were formed during the following decade, but as in the days of the monarchy, it was the authorities who rationed seat allocations for the favoured recipients. Here indeed is an example of those parliaments which Jalal Kishk, an Islamic critic of the Egptian military regime, has bitingly described as having been elected by nomination, or nominated by election.

To liquidate the Nasserist fiefs in the bureaucracy, the armed forces, and the security services, Sadat had need of friends and supporters. Among those whom he tried to draw to his camp were the Muslim Brethren who, as has been seen, had been harshly persecuted, and proscribed. As a young officer Sadat had had dealings with the Brethren and he was sympathetic to them. After 1970, though he did not legalize the Brethren, he yet allowed them elbow room to start their own publications and spread their views. Ironically, he was killed in October 1981 by an extremist offshoot of the Brethren, who believed that his regime was ungodly, and that he deserved to die. Sadat was succeeded by Air Vice Marshal Husni Mubarak, whom he had appointed Vice President in 1975.

In criticizing the military regime established after 1952, writers such as Kishk call the heirs to the monarchy Mamelukes or Pharaohs. In using such terms they wish to say that their rulers are self-chosen and self-appointed, and simply anointed by the customary large majorities registered in popular referenda. They also wish to indicate that the actions of these rulers, whether beneficent or not, are arbitrary and despotic, in that the ruled cannot question or modify them. The truth of such strictures emerges out of the record.

As has been seen, in Syria too, officers had taken matters into their own hands after 1949. Their actions led to the short-lived union with Egypt which grievously disappointed their dreams of an Arab union fashioned under their direction and according to their ideals. A military conspiracy in Syria in September 1961 ended the union with Egypt. The officers declared that they wanted to re-establish a civilian government. Elections were held in December and a civilian government was installed in office. It proved short-lived. In March and April 1962 new military *coups* took place pitting Nasserist and anti-Nasserist officers against each other. Deadlock ensued and a new civilian government was installed. At the end of the following July Nasserist officers acting in cooperation with Ba'thist officers executed another *coup*. These Ba'thist officers drew inspiration from the civilian Ba'th leaders Aflaq and Bitar, who were now again in favour of Nasser, and who had broken with their partner Akram Hourani, still opposed to Nasserism. The effects of this *coup* did not last very long, for on 8 March 1963, another group of Ba'thist officers executed yet another *coup*, which was to prove a turning point in the fortunes of Syria.

The authors of this *coup* had begun their association when they were young junior officers, stationed in Cairo during the short-lived union era, and where they formed a secret Military Committee. All five of them were devotees of Ba'thist principles, greatly disgruntled with the Nasserist regime, and with the condition of Syria before and after the union with Egypt. Three of them were Alawites and two Ismailis, that is to say they all belonged to small disfavoured minorities, looked down upon by the Sunnis as heretics, and held to be of no account by the Sunni urban notables who, for so long, had monopolized the wheeling and dealing which constituted Syrian politics. After the destruction of the union they, along with other Ba'thist officers, were cashiered by the Syrian command, composed of officers who had benefited from the overthrow of Nasser's regime and who were frightened not only of Nasserist, but also of Ba'thist plots. However, cashiered as they were, the members of the Military Committee continued to plot, and took part in partnership with the Nasserists in the *coup d'état* of 1962. In the *coup* of 8 March 1963 they also acted in partnership with Nasserist and other officers. The *coup* succeeded, the civilian government and its military protectors were arrested, and all power concentrated in a secret twenty-man National Council of the Revolutionary Command. The Council was composed of twelve Ba'thist officers, now reinstated, and eight Nasserist and unattached officers. The Nasserists, however, were soon shunted aside. They attempted to challenge the Ba'thists by force and organized riots in Damascus in May.

The Ba'thists suppressed the riots, shooting dead fifty rioters. They also arrested and tried Nasser loyalists among the civilians. Nasserist officers attempted a counterstroke in July with secret Egyptian help. They tried to occupy the radio station and the army headquarters. Hundreds of people were killed in the ensuing clash; twenty Nasserists were brought before military tribunals, sentenced to death on the spot, and immediately executed. Hundreds of Nasserists were also arrested and consigned to prison without trial.

The secret Military Committee was now close to controlling the armed forces and the Syrian state. As has been said, the Committee had five members. These were: the Alawite Muhammad Umran and his two fellow Alawites, Salah Jadid and Hafiz al-Asad; and two Ismailis, Abd al-Karim al-Jundi and Ahmad al-Mir. Umran was a Lieutenant-Colonel, Jadid and al-Mir Majors, and Asad and al-Jundi Captains. Aware of their lack of seniority and of belonging to sects considered heretical by the Sunnis, they co-opted a Sunni, Colonel Amin al-Hafiz, in a manner reminiscent of the Egyptian conspirators' co-optation of Naguib. When the *coup* of 8 March 1963 took place, he was military attaché in Buenos Aires. The successful conspirators had him recalled and gave him the Ministry of the Interior. During the attempted Nasserist *coup* the following July, the Minister illustrated himself by directing the defence of the radio station and the army headquarters, sub-machine-gun in hand. It is not recorded how many of the hundreds killed and wounded died at his valiant hands. The Minister of the Interior now became, in addition, Commander-in-Chief and President of the National Council of the Revolutionary Command. But he was not destined to remain long at the summit of power. Major Jadid and Captain Asad, now both Major-Generals, were not willing to allow Hafiz to become the supreme and unquestioned ruler of Syria. The period between the downfall of the Nasserists and February 1966 was one of obscure and ferocious intrigues and manœuvrings which ended by undermining the position of Hafiz and of Umran, who had become his ally, to the benefit of Jadid and Asad. These worked methodically and pertinaciously behind the scenes to control the army, the administration, and the Ba'th Party through their followers and creatures. On 22 February 1966 Hafiz's private residence was attacked by an armed band led by Salim Hatum, a Druse officer, and by Rif'at al-Asad, Hafiz al-Asad's brother. Hafiz and his bodyguards tried to fend off the attack, but had at length to surrender, after their ammunition became exhausted and a tank was brought up which demolished the house. Hafiz, Umran, and their supporters were taken to prison, the army was purged of those suspected of being loyal to the fallen rulers, and the civilian leadership of the Ba'th—Aflaq, Bitar, et al.—were deposed. Like Aflaq, Hafiz eventually found refuge with the Ba'th regime of Iraq, which seized power in 1968.

Umran, too, was eventually released from prison and went to Tripoli in Lebanon. He remained hopeful that he would be able to overthrow his two former colleagues in the Military Committee, so much so that about February 1972 he audaciously announced to Asad—by then President of the Republic—that he was coming back to Syria, thus clearly daring his former associate to arrest him. The response was immediate: on 4 March he was murdered in his Tripoli home, at whose orders exactly could not, of course, be ascertained.

In the event Umran's family did not blame Asad for his death and Umran's son, Najih, called on Asad some six months later and wept in his arms.[22]

The fate of two other members of the Military Committee may be quickly told. During the Six Day War, al-Mir was commander of the Golan front. He failed to stand up to the Israelis and fled on horseback. He was retired from the army. He was a supporter of Major-General Jadid, by then engaged in a deadly struggle with Major-General Asad. After his retirement from the army, al-Mir was given a seat on the National Command of the Ba'th Party. This was one of the central organisms of the regime, which any contender for power would have to control. Asad succeeded in removing al-Mir from the National Command and having him sent as ambassador to Spain. The other Ismaili member of the Military Committee was Abd al-Karim al-Jundi. After the *coup* of February 1966, he emerged as a supporter of Jadid, who had become the most powerful figure in the country, even though nominally he was only Assistant Secretary-General of the Syrian Ba'th. Jadid appointed Jundi as chief of internal security. In this office he became famous for the fear which he and his greatly expanded organization instilled in the people.

An army of petty informers was recruited, arbitrary arrests became frequent, and tales of torture, not hitherto common in Syria, contributed to an atmosphere of terror. Few dared to go about the capital after dark for fear of being stopped by the security police and taken away. People were even reluctant to leave the country because security agencies were known to confiscate empty houses. Much of this nastiness was linked with the name of Jundi.

In 1968 he was a clever man of thirty-six but with something not quite normal in his make-up, a penchant for cruelty which suggested that he was more of a nihilist than a radical socialist . . . Jundi treated jokes about himself as criminal offences. When he heard that a group of lawyers and other professional men had gossiped about him at a private gathering, he moved to arrest them, forcing several to flee on foot to Lebanon . . . [23]

The downfall of this man of terror came about in February 1969. By then Asad and Jundi's patron, Jadid, were at daggers drawn. Asad, who had become Defence Minister, together with his brother Rif'at, persuaded that Jadid was seeking to do away with them, themselves began pursuing Jadid's followers and driving them out of their positions. Jundi believed at one point that Rif'at was on the point of cornering him and committed suicide. We are told by Patrick Seale, 'on good authority', that, hearing of Jundi's violent end, Asad wept.

The fortunes of one other officer who took part in these bloody rites should also perhaps be recorded. Salim Hatum, as has been seen, led the assault on Hafiz's house in February 1966. Afterwards, considering that he had not been properly rewarded, and suspicious of the intentions of his fellow conspirators towards himself, he formed a secret Military Committee of his own, composed of fellow Druse, and headed by a Druse general, Fahd al-Sha'ir. They organized a *coup* the following

September which almost succeeded. Hatum captured Jadid, who was on a visit to the Druse Mountain. However, Asad, now Defence Minister, who was in Damascus, heard what was afoot and sent troops and aeroplanes to the Mountain. Sha'ir was captured, and Seale reports a story that 'he was made to get down on all fours like an animal and was ridden by his tormentors through dirty water'. Hatum fled to Jordan. When the Six Day War broke out he returned to Syria, possibly in order to take part in the war. Once in Syria, he was arrested and brought before a military court which confirmed the death sentence which it had earlier passed *in absentia*. While awaiting execution, Hatum was set upon by Jundi, who broke the condemned man's ribs.

After Amin al-Hafiz had been ousted, there were two contenders for power, the fellow Alawites Salah Jadid and Hafiz al-Asad. For the time being Jadid had the primacy, but rivalry between the two Major-Generals became increasingly sharp and, as has been seen, erupted into the open in a clash between the two sides. When, following this clash, Jundi killed himself, Jadid's faction lost a powerful supporter. Asad, furthermore, as Minister of Defence had been able to place his own followers in crucial posts in the armed forces. In November 1970, Asad moved to arrest Jadid and his supporters. The erstwhile comrade of the Military Committee now consigned his comrade to prison, where he remains to this day, while the young airforce captain of 1961 has been able, all these years, to keep his foothold on the slippery heights of power.

Since the first military *coup d'état* took place in 1949, Syrians have been the plaything of a handful of military officers engaged in deadly rivalries and murderous struggles. It is the whim of these officers which has decreed the social and economic arrangements which govern their lives, and it was their unfettered decisions which subjected Syrians to the hardships, the deprivations, and the bloodshed which wars entail. What is true of Syria is equally true of Iraq where, after 1958, a succession of military *coups d'état* again gave a handful of officers the unfettered power to bind and loose—unfettered, that is, except by fellow officers' conspiracies and fears thereof.

Brigadier Qasim's troops massacred the royal family on 14 July 1958, and the victorious conspirators took over the government. They were to be the harbingers of peace, prosperity, and Arab unity. Soon, however, dissension appeared, its ostensible cause lying in the leaders' differing views about the character of this unity. Qasim, now dubbed the Sole Leader, mistrusted schemes of Arab unity which would give Nasser the primacy. His distrust was all the greater that Nasser was, until the break-up of the United Arab Republic in 1961, his near neighbour in Syria. Qasim was determined not to allow Nasser or the Nasserists to dictate his policies and dominate his regime. The most prominent of his fellow conspirators, Colonel Abd al-Salam Arif, professed to be on the contrary a fervent Nasserist. The two fell out and Arif began to gather a following among fellow officers and the handful of Iraqi Ba'thists, all of whom strongly objected to Qasim's hostile stance towards Nasser. Qasim sought gradually to strip Arif of power and influence. Arif, who considered that he had played the crucial role in the events leading to the downfall of the monarchy, was naturally very resentful of these attempts to diminish and destroy his position. At a stormy meeting in Qasim's office he is alleged to have drawn his revolver—either to kill Qasim or to commit suicide. He was arrested on the spot, tried, and sentenced to prison. Qasim released his former associate in 1961, in what proved to be for him a fatal decision.

Also in 1959, other officers who also felt that Qasim was betraying the cause of Arab unity, and who resented Arif's downfall, conspired with the United Arab Republic to clinch an armed movement against Qasim. In the event, the army commander in the northern city of Mosul, Abd al-Wahhab al-Shawwaf, declared a rebellion which proved to be premature. He was killed by pro-Qasim forces, and his accomplices—like him, Qasim's comrades-in-arms—arrested, tried, and executed.

The events in Mosul were accompanied by a relatively new phenomenon: mobs on the rampage who were called in to support one side or the other and who indulged in indiscriminate slaughter. The phenomenon first appeared on the day the monarchy was overthrown. Arif occupied the broadcasting station in the capital and began inciting the mob to take to the streets and to attack the leading figures of the fallen regime. It was this mob which seized and mutilated the corpse of the Crown Prince, Abd al-Ilah, and strung up the remains at the gates of the Ministry of Defence. A few days later, a mob also caught in the street Nuri al-Saʻid, who was trying to flee disguised as a woman, murdered him, and drove a motor-car over his body, to leave it so horribly mangled as to be unrecognizable.

Shawwaf now called in Sunni Arab tribesmen from nearby in order to attack and kill left-wingers in the city, suspected of supporting Qasim. In retaliation, Kurdish tribesmen from the vicinity, together with Communist sympathizers, were let loose by Qasim in order to punish Shawwaf's supporters among the Mosulis. This new, frightful weapon was also used by Qasim in Baghdad. He allowed the formation of a so-called Popular Resistance Force, who became a law unto themselves, killing and terrorizing at their whim anyone they suspected of opposition to Qasim or of sympathy for the United Arab Republic and Nasser's brand of Arab unity.

Another explosion of mob violence also occurred in 1959, now, however, without the ruler's permission. The Communists, as small and as conspiratorial a group as the Baʻth, had been given licence by Qasim, in the Mosul affair, to attack, intimidate, and kill those suspected of working against him. They now thought to improve their position by showing the extent of their power. Kirkuk, the centre of the oil-bearing areas north of Baghdad, lay in Kurdish territory, but the city itself contained a very large Turkish-speaking community. On 14 July 1959 at the celebration of the first anniversary of the *coup d'état*, a mob of Kurds said to belong to the Kurdish Democratic Party, which was affiliated to the Iraqi Communist Party, fell upon the Turks of Kirkuk, indiscriminately killing, raping, and looting. Qasim took a dim view of the incident, now feeling able to dispense with the support of the Communists, and began methodically to clip their wings.

Qasim lasted until February 1963, when a conspiracy of

Ba'thist and Nasserist army officers cornered and killed him in his office in the Ministry of Defence. It is said that his murderers buried him in a shallow unmarked grave which dogs uncovered in order to eat his corpse. This horrified some fellahin nearby, who provided a coffin and gave the corpse a decent burial. This came to the ears of the secret police, who disinterred the corpse and threw it in the River Tigris.

Arif, whom Qasim had pardoned two years earlier, was part of the conspiracy and became President of the Republic. A Ba'thist officer, Ahmad Hasan al-Bakr, became Prime Minister. The new regime formed a National Guard which far exceeded in lawlessness the Popular Resistance Force. They were under the control of a civilian Ba'thist, Ali Salih al-Sa'di, Deputy Prime Minister. He was a man of thuggish and violent instincts who allowed the National Guards full scope in terrorizing the population. Sa'di's reign of lawlessness lasted for a summer and an autumn. His enemies within the Ba'th, which was now split, engineered his downfall in November, in co-operation with Arif and Nasserist officers. The Army was used to bring the National Guard to heel, and the fallen Sa'di had to go into exile. This *coup d'état* within the *coup d'état* strengthened Arif's hands, who proceeded to ban the Ba'th, expel the Ba'thist officers from the government, and govern with the help of Nasserist officers. One of these, Arif Abd al-Razzaq, Prime Minister in September 1965, feeling that the President was lukewarm in supporting a new Arab union under Nasser, attempted a *coup d'état*. He failed and fled to Egypt. He was replaced by the first civilian Prime Minister since 1958, Abd al-Rahman al-Bazzaz, an educator who had made, in 1952, a seminal contribution to the doctrine of Arab nationalism and its relation to Islam, which will be examined below. Bazzaz remained in office from September 1965 to August 1966. In April 1966, Abd al-Salam Arif was killed in a helicopter crash, and the Army officers appointed his brother Abd al-Rahman Arif, also an officer, to succeed him. Bazzaz now tried to settle amicably the Kurdish rebellion by conceding some of the Kurds' demands. He also had hopes of calling an election and establishing a representative assembly. Arif Abd al-Razzaq returned from Cairo to attempt a second *coup d'état*. He was foiled, but the officers in power, also believing that Bazzaz's Kurdish policy was harmful to Arab unity, required his dismissal. He was followed by an officer, Naji Talib, who held office until May 1967, when he had to go, following dissension among his fellow officers. They could not agree on a successor, and the President became also the Prime Minister. He gallantly bore the great burdens of his dual office until the middle of 1967, when he appointed another officer, Tahir Yahya, to be Prime Minister.

In July 1968 the Ba'thists came back with a vengeance. In 1966 Bakr had become the leader of the Iraqi Ba'th, with his fellow-Tikriti and kinsman Saddam Husayn as his second-in-command. The Ba'th had a following among the officers, but they were not strong enough to carry out a *coup* by themselves. They suborned two senior officers who were not Ba'this: the Commander of the Presidential Guard, Abd al-Rahman al-Da'ud, and the head of Army Intelligence, Abd al-Razzaq al-Nayif, by promising them high office should the *coup* succeed. On 17 July 1968 the *coup* was executed, Arif sent into exile, and his Prime Minister arrested on charges of corruption. Bakr became President, Nayif Prime Minister and Da'ud Minister of Defence. Thirteen days later another *coup* toppled Nayif and Da'ud. Nayif, emerging from a lunch with Bakr, was arrested by a group which Saddam Husayn led, and sent into exile. Da'ud was abroad and told not to return. In 1974 an attempt was made to murder Nayif in his London flat; four years later, another attempt was successful, and Nayif was gunned down in a London hotel. The Ba'thist *coup* of 30 July led immediately to the arrest, dismissal, and forced retirement of a very large number of officers and officials. In October the Revolutionary Command Council, now the sole authority in the country, announced the discovery of a plot against the regime. Among those arrested were a number of Jewish businessmen and some Muslim public figures, of whom the most prominent was Abd al-Rahman al-Bazzaz, the former Prime Minister, who had made his debut as an ideologue with a notable address to the Ba'th Club in Baghdad in 1952. The Jews were hanged, and

Bazzaz, accused of spying for Israel, was tortured while in detention and sentenced to fifteen years in prison. He was released in 1971 and died in 1972.

The Ba'th Party systematically and meticulously extended its control over all aspects of Iraqi life: over politics, society, and economy. The control was exerted by means of terror, the instrument of which was a pervasive network of secret police and intelligence services. Imitating the organizational pattern perfected by the Soviet Communist Party, the Ba'th likewise made sure that the population at large would observe its directives. Tentacular control was exercised through Party branches set up in neighbourhoods, in colleges and schools, in government offices, and army units.

Though the two *coups d'état* were carried out by army officers, the interpenetration of Army and Party made it possible in the space of a single decade for a civilian, Saddam Husayn, operating as the Vice-Chairman of the Revolutionary Command Council, to establish his own personal control over the Army and the state. Early on in the new regime, two Ba'thist senior officers who might have stood in Saddam Husayn's way were got rid of. General Hardan al-Tikriti, Minister of Defence and Deputy Commander-in-Chief, was dismissed in October 1970 and murdered in Kuwait the following March. His colleague and rival, General Salih Mahdi Ammash, was likewise dismissed in September 1971 and exiled to the embassy in Finland. Other Ba'thists of long standing were also done away with: Abd al-Karim al-Sheikhly, the Foreign Minister; Abd al-Karim Mustafa Nasrat, another Minister, were both murdered. So was Fuad al-Rikabi, leader of the Ba'th Party between 1952 and 1958. The only challenge to Saddam Husayn's dominance was made by a fellow-Ba'thist, Nazim Kzar, who was Chief of Internal Security and known for his brutality. In July 1973 he plotted to murder Bakr and Saddam Husayn. He took hostage two Ba'thist senior officers, the Ministers of Defence and of the Interior, and killed the Minister of Defence when his plot misfired and he tried to flee the country. He was caught and executed together with his accomplices.

By 1979 Saddam Husayn, then forty-two years old, had become the dominant figure in the Party and in the state. He had begun his political career in 1959 when, twenty-two years old and still at school, he took part in an abortive attempt to murder Brigadier Qasim. He escaped to Syria and then to Egypt, returning to Baghdad when the Ba'thists briefly seized power in 1963. Following their second seizure of power in 1968, Saddam Husayn steadily increased in power and prominence. He tried to settle the Kurdish rebellion by negotiations and then by force. Force proved unavailing since the Shah of Iran was supporting the Kurds, and the Iraqi Army was sustaining numerous casualties, which provoked unrest in the Shiite shrine cities, a great many of the casualties being Shiites. In 1975, Saddam did a deal with the Shah, whereby in return for frontier adjustments long desired by Iran, the Shah agreed to abandon the Kurds to their fate. The Kurdish rebellion collapsed, and the Kurds were left to the tender mercies of the Army, which was now able to kill and deport at will.

On 16 July 1979, Saddam Husayn succeeded Bakr as President of the Republic, the choice, it was declared, of the Revolutionary Command Council. It is, however, said that Bakr did not go willingly and that he was kept under house arrest. On 28 July, it was officially announced that 'a treacherous and lowly plot perpetrated by a gang disloyal to the party and revolution' had been discovered. On 8 August, it was announced that the death sentence had been carried out against 'a handful of criminals who conspired against the Party, the revolution, and the country's higher interests', and that 'the execution was attended by the Comrade President, members of the Special Court, the Comrade members of the Investigation Committee, and leading Party cadres'. The confession of the chief culprit, which incriminated his Party comrades, was videotaped and shown at Party meetings. This culprit was the Secretary of the Revolutionary Command Council, who was killed along with all his family. The other members of the Command Council who perished on this occasion included the Minister of Kurdish Affairs, the Deputy Prime Minister (who had been appointed to this post on 17 July), the Minister of

Education, the Minister of Industries, the Director-General of the President's Office, and the Head of the Vice-President's Office. Many others were also killed on this occasion.

This, then, is the dismal picture which many parts of the Middle East present to the onlooker: countries which have been confiscated by a handful of conspirators, usually military, who come to control their often very large resources and populations and use them at their whim. This is, moreover, the case not only in the countries mentioned above: Egypt, Syria, and Iraq, which are the most important Arab states in the Middle East, but also in the two less populous and more peripheral countries: Libya, where a military *coup d'état* took place in 1969 and where the original conspirators are still in power; the Sudan, which began at independence in 1956 as a constitutional and parliamentary Republic, but which has suffered a succession of military *coups d'état* beginning in 1958, whose leaders have led the country to poverty and civil war between the Muslim north and the Christian and pagan south; in Algeria, where the officers of the Front de Libération Nationale have been continuously in power since the French government handed the country over to them in 1962; and in Tunisia where an army officer in 1987 deposed his civilian predecessor who had himself established a one-party state and ruled unchallenged since the French abandoned their protectorate in 1956. Iran, too, as will be seen, lives under the unfettered rule of a small, albeit civilian, group.

Such seems to be the outcome of nearly two centuries' contact with Western political institutions and ideals, which one generation after another has passionately wished to emulate, and tirelessly worked to make a reality in the Middle East. In 1969, a former Syrian Ba'thist who served as a minister and an ambassador, and who was then living in exile, Sami al-Jundi, published a short book of political recollections, in which he looked back with bitterness on the high hopes with which as a schoolboy he had joined the Party—hopes of freedom and prosperity, of a society cleansed of corruption and of outmoded traditions:

We thought that the epochs of decadence had come to an end with our predecessors among the politicians, and that we were the glorious beginning of a new civilization, when in fact we were the last exemplars of backwardness, and a desolating expression of it.

We wanted to be a resurrection [*ba'th*] of signal deeds and heroism, but what was resurrected through us—when we came to power—was no more than the period of the Mamelukes.

'Who would have believed during the school-year 1940–1' Jundi begins his recollections by asking,

that the Ba'th would end in all this mockery? Who among us would have thought that the word, Ba'thist, would become an accusation which some of us would rebut with bitter scorn? . . .
Our sacrifices, our youth wasted on the roads among the people, our dreams and our faith, were they then all a mockery?[24]

Jundi's rueful words do not apply only to the Ba'thist regimes in Syria or in Iraq, particularly arbitrary, oppressive, and blood-soaked as they have proved to be. Could not similar thoughts have revolved in Nuri al-Sa'id's head as, disguised in a woman's attire, he was being lynched by the Baghdad mob on a July day in 1958; or in that of his brother-in-law, Ja'far al-Askari, as he, the Minister of Defence, was being murdered on a roadside outside Baghdad on that day in October 1936 at the behest of one of his generals; or in that of Abd al-Hakim Amir, as he lay dying from poison in the bathroom of the rest-house outside Cairo, fearing and hating the comrade of his youth Jamal Abd al-Nasser; or in that of Abd al-Rahman al-Bazzaz as he helplessly submitted to the torturer's ministrations in a Ba'thist dungeon?

6. The Politics of Radical Islam

The address which Abd al-Rahman al-Bazzaz gave to the Ba'th Club in Baghdad in 1952 was entitled 'Islam and Arab Nation-alism'.[25] In it he dealt with an issue which went to the heart of the ideology of Arab nationalism. As has been seen, one important point on which Sati' al-Husri insisted was the primacy of Arab nationalism over Islam, that what counted in nationalism was a common language and a common history, rather than a common religion. Such an argument, however, must face many difficulties. The Arab nation which Husri had in mind was overwhelmingly Muslim, and strongly attached to the beliefs, traditions, and mores of a religion which had lasted, at the time he was writing, for over thirteen hundred years. Was it likely that they would abandon their religion for a new-fangled ideology? Or, if they were attracted to Arab national-ism, how would they resolve the conflict between the new loyalty and the traditional one? Bazzaz's achievement was to demonstrate that opposition between Islam and Arab national-ism was illusory and conflict between them highly unnecessary. Islam was indeed a universal religion suitable for all peoples, but it 'is undoubtedly a religion first revealed to the Arabs themselves; in this sense it is their own special religion. The Prophet is from them, the Koran is in their language; Islam retained many of their previous customs, adopting and polish-ing the best of them'. Muhammad's was indeed a universal Message, but its first effect was to 'revive' and 'resurrect' the whole Arab nation. There is, in brief, absolutely no divorcing Arabism from Islam: the Prophet was an Arab, he came out of an Arab environment, and his revelation was imparted in the Arabic language. Instead of disparaging Arab society before Islam in order, mistakenly, to enhance the glory of Islam, Bazzaz argued, Muslims should recognize that the society must have possessed a high and sophisticated civilization for the Prophet to come out of it. Such an argument went, of course, against Islamic tradition in which Arab society before Islam was *jahiliyya*, a Time of Ignorance, sunk in paganism and superstition. But this Islamically suspect argument apart, Bazzaz did succeed in effecting a tight link between Arabism and Islam: 'the Muslim Arab, when he exalts his heroes, partakes of two emotions, that of the pious Muslim and that of the proud nationalist'. Bazzaz went even further in putting forth a daring argument in comparing the primacy of the Arabs within Islam to the primacy of the Communist Party of the Soviet Union within the Communist world.

If the ideology of Arab nationalism was to attract the public at large in the Arabic-speaking world, then, it had to be shown that it was not in conflict, or incompatible, with Islam. To have put together an argument establishing this point was Bazzaz's achievement. It is significant that the necessity of doing this was also realized by non-Muslim proponents of Arab nationalism, for whom this ideology had originally seemed to offer an escape from sectarian divisions, and from the inferior position to which non-Muslims are confined in a Muslim polity. Thus we see Professor Constantine Zurayq of the American University of Beirut, a believer in the necessity of Arab nationalism, calling upon all Arab nationalists, regardless of their religion, to venerate the Prophet's memory. In a speech given in 1938 on 'the anniversary of the birth of the noble Arab Prophet', Zurayq declares:

Today . . . we need more than ever, in our national struggle, to have leaders who may draw from the personality of the Arab Prophet strength of conviction and firmness of belief. In this way they may engage, with courage and daring, in the struggle for their principles, calmly facing persecution and scorning obstacles, that they may inspire in the breasts of those sons of the Arab nation who are around them the spirit of sacrifice and devotion, carrying them on to the straight path which leads to the new life. This is the spiritual message contained in the anniversary of the Arabian Prophet's birth which is addressed to our present national life. It is for them, in spite of their different tendencies and their diverse religions and sects, that the Arab nationalists must honour the memory of Muhammad ibn Abdullah, the Prophet of Islam, the unifier of the Arabs, the man of principle and conviction.[26]

Aflaq was of the same mind as Zurayq. In a lecture of 1943 he also declared it incumbent on every Arab nationalist to venerate the Prophet who, through his message, was the first to unite the Arabs. After he took refuge from the Syrian Ba'th with the Iraqi Ba'th in Baghdad, Aflaq, following his view to its logical conclusion converted to Islam.

These are attempts, it is clear, to square Arab nationalism, a new and unfamiliar ideology, with the Islamic belief to which the mass was traditionally and deeply attached. However, at the very time when a nationalist ideology was being worked out, intellectual and popular developments within Islam itself were in train which were to culminate in an influential and politically significant Islamic ideology. As has been seen, Afghani had many far-reaching criticisms—both exoteric and esoteric—to make of the condition of Islam in his day. His exoteric critique, namely that Islam was stagnant, full of superstitions and thus an easy victim to alien, i.e. European, currents of thought, was continued and expanded by his disciple Muhammad Abduh—particularly when he moved from his master's influence and became an influential personality on his own—a man whose views were eagerly solicited by many educated Muslims and received respectfully. One of Muhammad Abduh's disciples, certainly the most eminent one, more eminent, in fact, than Abduh himself, was Muhammad Rashid Rida (1865–1935). A native of Tripoli in Syria, he came to Egypt, attracted by what he could gather from afar about Afghani's and Abduh's teaching. In Cairo, he sat at Abduh's feet until the Mufti's death in 1905. He also started a periodical, *al-Manar*, which appeared monthly from 1901 until his death in 1935, Rashid Rida himself largely writing each bulky number. *Al-Manar*, we may say, was dedicated to the enterprise of regenerating Islam, of making it able to cope with a modern world in which civilization, technology, and political power resided outside the Muslim world. Rashid Rida's recipe for this regeneration was that Muslims should go back to the ways of the ancestors, the *salaf*, who practised a pure and rational Islam, free from superstition, and having within it the resources to establish a just and prosperous polity. The only answer to present difficulties, distresses, and humiliations would be to adopt the way of the Prophet and his immediate successors, to follow the 'straight path' indicated by the Koran, the traditions, and the Prophet's example.

This *salafi* doctrine became very influential in many parts of the Muslim world. Rashid Rida and his *Manar* had an audience which extended far beyond the Arab world, as the letters and queries addressed to him by Muslim readers in India, the Far East, Africa, and the Americas testify. Rashid Rida's call to a purified Islam which would go back to its pristine state was not a specifically political programme, nor did Rashid Rida attempt to establish or to lead a political movement. Another much less learned and much more activist figure, the Egyptian Hasan al-Banna (1906–49), who was very much influenced by Rashid Rida's writings, did attempt to do so. Banna, a provincial schoolmaster, observed the social and economic condition of the urban poor, usually migrants from the countryside in search of a livelihood, and concluded that these were helpless and abandoned multitudes whose miserable living conditions and spiritual deprivation had reduced them to a kind of social dust. Neither the government nor the official religious establishment paid any heed to these lost souls or did anything to alleviate their miserable condition. In 1928 Banna established the Muslim Brotherhood as an association devoted to the amelioration of the condition of the poor and the deprived, by teaching them ways of self-help, by restoring their self-respect, and tending to their spiritual welfare. Banna's activity was built on the premise that the way to salvation lay in a return to the Koran and in the practice of solidarity and mutual help according to the maxim that all believers are brethren. The Muslim Brotherhood obviously answered a crying need, for during the 1930s it grew by leaps and bounds; it established an extensive network of branches and of co-operative enterprises, and its teaching found favour not only among the poor and the illiterate, but also, increasingly, among the educated and within the professions. It was what may be called a grass-roots movement, in that it was neither the outcome of official action, patronage, or inspiration, nor was it indebted in any way to the Westernized political leaders who played the political game in Cairo. Exact numbers of adherents do not exist, but an idea of the order of magnitude involved may be gained from Richard P. Mitchell's estimate, in his study, *The Society of the Muslim Brothers* (1969), that in 1946–8 the Society had some half a million members and another half a million sympathizers.

The phenomenal growth of his Society led Banna, the Supreme Guide, according to his official title, to aspire to play a role in politics. Nor is this surprising. If Muslim society was in danger, if it was now wide open to foreign corruptions and to foreign domination, whether blatant or insidious, then the only remedy was to make the government of Egypt fully Islamic, to make the Koran its constitution. Only thus would the Muslims be protected from foreign aggression and corruption. This is what has come to be known as Islamic fundamentalism, nor is it, in this respect at any rate, any different from the teaching of Ayatollah Khomeini with which this term has come to be generally associated. In sum, from being a movement devoted to promoting the material and spiritual welfare of Muslims, the Muslim Brethren became an increasingly activist and radical political movement. Between 1939 and 1945 war conditions and British control over Egyptian political activities allowed no scope for Banna to intervene in politics. The years which immediately followed the end of the war witnessed a great deal of turmoil, where political leaders vied with one another in denouncing the British presence in Egypt, and indulging in mutual accusations and recriminations. Student demonstrations, particularly in Cairo, accompanied and materially increased the political turmoil. The Brethren played their part in increasing and exacerbating this turmoil. Banna could not only call upon his mass following to demonstrate, protest, and riot, he could also make use of a weapon which was at his disposal. It is not exactly known when Banna established the Secret Apparatus, a clandestine terrorist organization, which would be used to intimidate and to eliminate opponents. But in the years between 1945 and 1949 the Apparatus attempted, and sometimes successfully carried out, the assassination of various political figures, and of judges who had delivered incriminating verdicts in trials involving Brethren accused of terrorism. When Egypt joined the Palestine war in May 1948, the Government declared martial law, and used its new powers to suppress the Brethren. On 8 December 1948, Nuqrashi, the Prime Minister acting in his capacity as Military Governor under martial law, decreed the dissolution of the Society, an immediate ban on its activities, and the confiscation of all its funds and property. Twenty days later, Nuqrashi was murdered by a Muslim Brother; a month or so later, on 12 February 1949, Banna himself died at the hands of unknown murderers. It was rumoured at the time that the instigation came from Faruq, or possibly from Nuqrashi's successor who himself survived an abortive attempt on his life the following May.

It is clear from all this that the success of Banna's political programme, namely, the establishment of an Islamic regime where all Western-inspired legal codes and political institutions would be abolished, was to be brought about by violence. There was no doubt about what Banna thought had to be destroyed, but the lineaments of the Islamic regime which would replace it were sketchy in the extreme. Violence indeed continued to be the hallmark of the political activity of the Brethren after Banna's death. Following Faruq's downfall in July 1952, the Brethren were in hopes that the Free Officers would take steps to realize Banna's vision, especially as some of the more important of these Officers had had connections or sympathies with the movement. These hopes were doomed to failure. Nasser, who emerged as the effective leader of the *coup d'état*, had no intention of allowing any organization to dictate his policies. The Brethren were disappointed, and one outcome of their disappointment was an intrigue against Nasser with General Naguib, who was the President of the newly proclaimed Republic. Another was an attempt to assassinate Nasser while he was giving a speech before a large crowd in Alexandria in October 1954. Severe repression fell on the Brethren. Hundreds were arrested and brought to trial before a newly established tribunal, the People's Court, composed of three Free Officers including Anwar al-Sadat. Hundreds of Brethren, civil and military, were tried, six were executed, and ten given various prison sentences. In addition, some thousands were put in internment camps, and all activity by the Brethren was proscribed and suppressed.

Banna, though he had been very much influenced by Rashid Rida, *al-Manar*, and *salafi* ideas, was not himself a systematic or original thinker. His talents lay more in exhorting, in organizing and in political negotiation and secret action—in which he finally came a cropper. One successor of his, Sayyid Qutb, was a talented and persuasive writer, endowed with a sharp analytical mind which enabled him to establish clear and well-formed distinctions between the organizing ideas of Islam and those of Western civilization. Qutb had started out as a literary man who wrote fiction and literary criticism, far removed from the ideas of Banna and the Muslim Brethren. He was invited, as a promising young writer, to visit the United States. His short visit effected in him a spiritual and intellectual revolution. He felt a great revulsion for the libertinism and godlessness which he came to believe were the hallmark of the American, and generally Western, mode of life. When he returned to Egypt he became one of Banna's followers and a very articulate exponent of the Islamic world-view and of its foundations in the Revelation. He was among those detained when Nasser suppressed and proscribed the Brethren, and remained in internment until 1964. The year following his release, he published a small book, part of a much larger work of Koranic exegesis. The book, entitled *Signposts on the Road*, so angered the authorities that they immediately published a rejoinder, *Signposts on the Road of Treason and Reaction*. He was arrested shortly afterwards, accused of conspiring to assassinate Nasser and other figures, sentenced to death, and executed in 1966.

The central organizing idea of *Signposts on the Road* is that of *jahiliyya*, i.e. the idolatry, the 'ignorance', from which Islam came to deliver the world. For Qutb, both the Western world and its Communist antagonist are sunk in idolatry, both denying the sovereignty of God, rebelling against His commands, and preferring to a godly life the pursuit of animal satisfactions. Qutb goes further: *jahiliyya* is now present not only in the capitalist West and the Communist East, it has also infected the world of Islam:

All that is around us is *jahiliyya*. Peoples' imaginings, their beliefs, customs, and traditions, the sources of their culture, their art and literature, their laws and statutes, much even of what we take to be Islamic culture, Islamic authorities, Islamic philosophy, Islamic thought: all this too is of the making of this *jahiliyya*.[27]

For Qutb it is above all Muslim rulers, infected with the spirit of idolatry, who are responsible for much of the evil. They are apostates from whose deadly clutch Muslim society has to be saved. Qutb's, then, was a much more radical and activist doctrine than Banna ever formulated, and it proved to be the inspiration which, during the following decade, drove small groups, offshoots of the Muslim Brothers, to organize conspiracies against personalities and institutions of what they believed to be an apostate regime. In 1974, a group tried to take over the Technical Military Academy as a prelude to an attack on notabilities of the regime assembled to hear a speech by Sadat. They were foiled, but a great number were killed and wounded. The two principal ringleaders were captured, sentenced to death, and executed. In 1977, another similar group kidnapped a well-known divine, who had been Minister of Pious Foundations, in order to force the Government to release some of their members who had been arrested. The Government refused to give in, and the divine was murdered. The Government unleashed a man-hunt in which hundreds were arrested and tried. Five of the top leaders were sentenced to death and executed.

The most spectacular exploit by these groups was the murder of Sadat while he was reviewing his troops in October 1981 on the anniversary of the Egyptian crossing of the Suez Canal eight years before. The perpetrator was Lieutenant Khalid al-Islambuli who was seized on the spot with his accomplices. In the subsequent trial, Islambuli proclaimed that he had killed the Pharaoh—who in the Koran stands as the epitome of tyranny—and that he was proud of having killed the unbeliever. Islambuli and his fellow conspirators drew their inspiration from the teachings of an engineer, Abd al-Salam Faraj, who was arrested, tried, and condemned to death along with the actual perpetrators of Sadat's murder. Faraj's teachings are contained in a short piece of writing, *The Forgotten Obligation*. In this Faraj argued that it was a religious obligation, as incumbent on the believer as the obligation to pray, or to fast, or to go on pilgrimage, to kill a Muslim ruler whose actions show that he collaborates with, or is on the side of, the unbelievers in their war against Islam. Such a ruler has thus become an apostate and his killing is lawful. Faraj's teaching is the logical culmination of Qutb's doctrine.

Sadat's murder was followed by an attempted uprising by the fundamentalists in upper Egypt who seem to have disposed of money and arms. The uprising was snuffed out, and the hope of esablishing an Egyptian polity solely governed by the *shari'a* remains unrealized. Even more desperate attempts were made by the Muslim Brothers to destroy the Ba'th regime in Syria. This regime was controlled by officers who belonged to sects which the Sunnis considered to be heretical, whether Druse or Ismaili or Alawite, and it was the Alawites, as has been seen, who in the end prevailed. A predominantly Sunni population did not easily accept being ruled by heretical upstarts whose yoke, furthermore, lay heavy on them. In 1964, there were clashes between Sunnis and Alawis, and riots erupted in the predominantly Sunni city of Homs. Loudspeakers mounted on minarets spread the war-cry, 'Islam or the Ba'th', and government troops were fired on from a mosque. The army responded by shelling and destroying the mosque.

After 1966, the Ba'thist regime was fully Alawite. Its sectarian character, the harshness of its rule, its incompetence in managing the economy, and its corruption kept the Sunni population antagonized and provided fertile ground for the Syrian Muslim Brothers, who became very active, indeed belligerent in challenging the regime. Assassinations and attempted assassinations followed one another. High officials, Ba'thist leaders, and Soviet advisers were the target. These culminated in an attempt to kill the President in 1980. Terrible retribution followed. Two units of the Defence Companies, commanded by Asad's brother Rif'at, were assembled one morning at dawn and instructed by his deputy and son-in-law to attack a prison in Palmyra in the Syrian desert where Muslim Brothers were held prisoner. On arrival, the troops were let loose on the defenceless prisoners, 500 of them, whom they murdered to a man.

In their clash with the Ba'thist Government, the Muslim Brothers were able to enlist the sympathies of whole quarters, or even cities. Also in 1980, in March, demonstrators sympathetic to the Brothers attacked barracks and Ba'th Party offices in Jisr al-Shughur, a small town near Aleppo. The government sent troops by helicopter to engage the demonstrators, and 200 were killed. A few days afterwards, troops were sent to attack whole quarters in the city of Aleppo, and put an end to rebellion by the Muslim Brothers.

Standing in the turret of his tank, the divisional commander, General Shafiq Fayadh, told the townspeople that he was prepared to kill a thousand men a day to rid the city of the vermin of the Muslim Brothers. His division stayed in Aleppo for a whole year, with a tank in almost every street.[28]

The following August, Aleppo, and in April 1981, Hama, received another visitation from the troops, who arrested at random and shot out-of-hand scores of males over the age of 14.

The culmination of the Brothers' war against the Ba'th was an uprising in the city of Hama in February 1982. The uprising led to a three-week battle in which 12,000 troops with tanks and helicopters besieged and hunted down the rebels, in operations which led to the destruction of about a third of the city, and the death of between ten and twenty thousand inhabitants. Happily, the authorities resolved to rebuild the town, and provided generous funds for the project. As Patrick Seale describes:

Heavily damaged old quarters were bulldozed away, roads were cut through where once no car could pass, squares and gardens were laid out. The whole of Hama was reshaped on a grand scale, with ring-roads and roundabouts serving entirely new quarters furnished with schools, clinics, playgrounds, and shopping malls. Among major public buildings put up after the rising were a 230-bed hospital, a cultural centre, a girls' sport institute and teacher-training college, a central market of oriental design, headquarters buildings for the Peasants' Union and the Federation of teachers and engineers, and a sports centre of outrageously ambitious proportions complete with Olympic-sized swimming pool. On Asad's orders, the state funded the construction of two large mosques to make up for those destroyed in the fighting as well as a Catholic church as large as a cathedral. Among the revolutionary changes was the introduction of mixed bathing in 1983 and the first college dormitory in the whole of Syria to house both male and female students. By then the Sporting Club had some eighty girl members and in 1985 Hama girls were the national ping-pong champions.[29]

Apart from demonstrating in this manner the beneficent cunning of history, the Muslim Brothers have not, so far, been able to realize the dream of a Muslim polity regulated exclusively by the Koran and the *shari'a*. The dream is, in effect, a mirror image of the ideology of Arab nationalism, to which it is so adamantly opposed. Just as the quest for a unified state to include all Arabs—Husri's and Aflaq's dream—seems to have led only to a great deal of bitter inter-Arab conflict, so the quest for a true Islamic society by the followers of Banna and Qutb has only issued in much violence in Egypt and Syria, and has not enhanced feelings of Muslim brotherhood and solidarity. In the Sudan, too, the ideology of the Muslim Brothers, adopted in 1983 by the military ruler, Ja'far al-Numeiri, who came to power in a *coup d'état* in 1969, was a contributory cause of the civil war between the Muslim North and the pagan and Christian South (comprising over a quarter of the population) which has been raging ever since. In September 1983, perhaps needing to attract the support of the Muslim Brothers when he was facing mounting discontent over his mismanagement of the economy and increasingly arbitrary rule, Numeiri suddenly announced that the *shari'a* would be the law of the land. Alcohol was forbidden, and about $5 million worth of beer, spirits, and wine were supposed to be poured out into the river and the sewers (in fact, it is said, 90 per cent of the drink was appopriated by the officials and soldiers responsible for destroying it), and amputation of limbs and other Koranic punishments instituted. It was declared that the law would not be applied to non-Muslims, but this was not sufficient to allay the great disquiet which Numeiri's sudden action aroused in the South. His downfall in 1985 and the policies of his successors—two leaders of a military *coup d'état* and a civilian government in the interval between the two—has done nothing to alleviate this.

Assuming that Banna or his activist followers in Egypt and Syria had obtained their wish and decreed a fully Islamic polity, the question arises whether the *shari'a* would be adequate for the governance of complex and sophisticated societies which, in addition, cannot disregard or be isolated from a world where the norms of civilization, scientific, and technological advance, together with political and military power are all found outside the world of Islam.

The matter was, to some extent, put to the test by a figure very different from Banna, Qutb, and their ideological progeny, and in a Muslim country whose traditions and experience were somewhat removed from that of Egypt and Syria. The figure in question was Ayatollah Ruhallah Khomeini (1902–89) who founded the Islamic Republic of Iran in 1979 and was practically its sole ruler for the following decade. Khomeini was not a lay figure like Banna or Qutb, but a Shiite divine recognized by his fellow divines as having attained a proficiency in the religious sciences entitling him to the honorific appellation of Ayatollah, which signifies a divine sign or miracle. The history of Shiism as the religion of a small minority, long devoid of political power, and subject to occasional persecution, meant that their divines had a position quite different from that of their Sunni counterparts. Sunni divines had generally been under the authority and control of the ruler, and held only a subordinate position in the ruling institution, taught to do the ruler's bidding and resignedly to acquiesce in his ways, however repugnant they might be.

Shiite divines, on the other hand, came to be considered as leaders of the community, to whom the faithful looked for authoritative guidance in matters of religion and law. A ruling attributed to one of the Imam Ali's descendants which has been accepted as authentic by the religious authorities declared that Shiites must shun recourse to a ruler or a judge whose power is illegitimate—because he does not recognize that Ali and his designated descendants are the only legitimate Imams of the Muslims. Instead they should seek out and accept as a judge one versed in the traditions of the Shiites and who knows their laws and traditions: 'For I appoint him as judge over you'. Shiite divines thus came to have a position very similar to that of the rabbis who, after the destruction of the Jewish state at the Romans' hands, kept together and preserved the cohesion and continuity of Jewish communities everywhere. At the beginning of the sixteenth century, Shah Ismail, the founder of the Safavid dynasty, made Shiism the state religion of Persia, which thus became the first (and has remained so far the only) Shiite state in the world. The Safavids claimed descent from the Imam Ali, and the Shiite divines accepted their title to rule. Even when they were succeeded by rulers who could claim no such descent, the divines never disputed the absolute title to rule of a monarch who accepted the tenets of Shiism. As a well-known divine put it in the last quarter of the nineteenth century:

Absolute kingship is the expression of God's power and belongs to the king who is the shadow of God on earth, in the same manner that the *ulama* are the expression of the religion. Thus they are both successors to the Hidden Imam, may God accelerate his advent. The unity of the *ulama* and the king is derived from the fact that they are both fruits of the same tree.[30]

This statement articulates and summarizes what, in Persian political experience since the sixteenth century, may be called not so much the division, as the duality of power, between ruler and divine. The duality was preserved and enhanced in the Persian Shiite state for three important reasons. In the first place, the faithful continued to recognize as incumbent upon them the payment directly to the divines of *khums al-Imam*, i.e. the fifth of commercial and agricultural profits, and men of religion continued thus to dispose of large incomes for the building and upkeep of mosques and religious colleges, and to spend on charity. In the second place, the authority of the divines was greatly enhanced as an outcome of a controversy between *akhbaris* and *usulis* during the eighteenth century. *Akhbaris* contended that religious authorities were bound to follow in their rulings the Koran and the Traditions of the Prophet and the Imams, while the *usulis* contended that authorities were free to use their reason and judgement in developing—creatively, so to speak—the jurisprudence which had come down to them. *Usulis* eventually won out, and this very greatly raised the position of the divines and very much increased the authority of each ayatollah over those of the faithful who recognized his primacy. An ayatollah so recognized became for his followers a *marjaʿ al-taqlid*, i.e. one whose authoritative rulings in matters of religion it is incumbent on them to follow. In the third place, two pre-eminent centres of Shiite religious learning, the shrine cities of Najaf and Karbala, were from the seventeenth century onwards under Ottoman rule, and divines who lived and taught in them were outside the reach of Persian rulers.

All this is to say that the divines in Shiite society enjoyed a popular standing and authority much higher than that enjoyed by their Sunni equivalents. When, as happened particularly during the Tobacco Protest, or during the events leading shortly afterwards to the Constitutional Revolution of 1906, divines were moved to attack the monarch and his policies, they felt able to do so, and a wide popular audience was receptive to their strictures, however harsh they were. Reza Shah Pahlavi, once in power, was determined that he would not suffer from the same criticisms and attacks endured by his Qajar predecessors, and took care to intimidate the divines and keep them under a strict control. His son followed in his footsteps.

However, one particular divine was courageous enough to speak out and criticize the Shah and his policies, and obstinate enough to go on doing so in the face of determined attempts by the authorities to silence him. Khomeini had been trained in the regular religious curriculum, and when he began denouncing the regime in the early 1960s for its ungodly policies and for its links with Israel and the United States, he was well-known and respected as an authority and a *marjaʿ* in the shrine city of Qum. In doing so, Khomeini was following in the footsteps of his fellow divines who had attacked the policies and behaviour of Nasser al-Din Shah and his son Muzaffar al-Din sixty and seventy years before. More specifically, he was not, at that point, questioning the institution of the monarchy, only denouncing the misdeeds of a particular monarch.

After many clashes with the authorities between 1962 and 1964, when the Shah's troops attacked his followers in Qum and elsewhere, inflicting a large number of casualties, and when Khomeini himself was arrested and kept for a year in confinement, he was finally sent into exile in Turkey. In 1965 the Iranian authorities agreed that he should take up residence in Najaf in Iraq. This is where he remained until October 1978, when he went to France, following pressure by the Iranian Government which led Iraq to terminate his stay in Najaf. From France, where he lived in a sympathizer's villa at Neauphle-le-Château near Paris, Khomeini returned in triumph to Tehran on 1 February 1979, to inaugurate the Islamic Republic. Its new institutions were designed to create a regime based entirely on the prescriptions of the *shara'i* and thus to replicate, in the view of its founders, the Islamic polity as it had existed at the time of the Prophet.

It was during his Najaf years that Khomeini put together a new and original theory of Islamic government, the like of which had never been heard before whether in Sunni or Shiite Islam. The theory was presented in a series of lectures to an audience of religious students between 21 January and 8 February 1970, and soon thereafter revealed to the world in a book, *Islamic Government*, the text of which had been recorded by one student direct from the Ayatollah's own discourse.

If it can be said that the relation between rule and religion in Islam turned out to be a kind of caesaropapism, where the ruler has the sole ultimate power over all his subjects, including men of religion, then in these lectures Khomeini has reversed this traditional and long-accepted order of things, and has given the man of religion the ultimate authority in all matters, whether religious, political, or military. The manner in which Khomeini conducts his argument and reaches his conclusion is very much the manner to which religious scholars are traditionally accustomed. Between this traditional mode, however, and the very radical conclusion to which the argument leads, the disparity is very striking indeed.

Khomeini argues that the sole legislator for humanity is God, who has revealed his commandments through the Prophet. During his lifetime the Prophet exercised government and executed the laws. To the Prophet there had to be a successor who would discharge the same functions. This, the only legitimate successor, designated by the Prophet, was the Imam Ali who exercised rule over the Muslims. Ali having gone, who, in the absence of the Hidden Imam, will then succeed to the Prophet's and to Ali's functions? Here Khomeini quotes a Tradition which plays a crucial role in his argument. In this Tradition Ali relates that the Prophet said: 'O God! Have mercy on those that succeed me.'

He repeated this twice and was then asked: 'O Messenger of God, who are those that succeed you?' He replied, 'They are those that come after me, transmit my traditions and practice, and teach them to the people after me.'[31]

We know, Khomeini argues, that Ali was the designated successor of the Prophet, and we know what in his case successorship entailed—namely, governance of the Muslim polity. The successors of whom the Prophet speaks in this Tradition, those who 'transmit my traditions and practice', it must follow, will have, as successors, the same functions as the Imam Ali whom the Prophet also designated as his successor. Furthermore, the governance of Muslims is governance according to law, and thus requires that those in charge of governance should be knowledgeable in the law. Who better than the learned and just divine is familiar with God's law? A person having these attributes, declares Khomeini, will 'possess the same authority' as the Prophet, and everyone will have a duty to obey him. Khomeini clinches his argument thus:

Islamic government is a government of law, not the arbitrary rule of an individual over the people or the domination of a group of individuals over the whole people. If the ruler is unacquainted with the contents of the law, he is not fit to rule; for if he follows the legal pronouncements of others, his power to govern will be impaired, but if, on the other hand, he does not follow such guidance, he will be unable to rule correctly and implement the laws of Islam . . . If the ruler adheres to Islam, he must necessarily submit to the [learned divine], asking him about the laws and ordinances of Islam in order to implement them. This being the case, the true rulers are the [learned divines] themselves, and rulership ought officially to be theirs, to apply to them, not to those who are obliged to follow the guidance of the [learned divines] on account of their own ignorance of the law.[32]

This conclusion Khomeini establishes to his own satisfaction. The conclusion, such as it is, however, sets at naught thirteen centuries of Muslim political experience in which no single instance exists of religious jurists exercising rule or acknowledged to be the only Muslims entitled to rule over other Muslims. Khomeini's, we may think, is a bookish dream which, for its crucial argument, depends on an absurdly wide interpretation of what is entailed by the Prophet's description of those who transmit his tradition and practise as his 'successors'.

When Khomeini's lectures were published no one took much notice. However, exactly nine years later he was back in Iran, received by its masses with rapturous welcome, and destined to be its absolute ruler for over ten years, until his death in 1989. The bookish dream of a divine becoming the ruler of a Muslim polity had been realized. It was realized in consequence of the Shah's impolicy and foolish ambitions—he claimed that he would make Iran the sixth great power in the world; in consequence as well as the flood of oil-money with which he drowned Iran after the OPEC coup of 1973, and which set up severe strains in the fabric of society, particularly urban society; in consequence, again, of the availability of cheap radio cassettes by which Khomeini's message was sown broadcast all over Iran by willing helpers, many of whom later fell victim to what Khomeini called the Islamic revolution; in consequence above all, perhaps, of the Shah's hesitant and infirm manner in dealing with the onset of trouble in the summer and autumn of 1978, and of the curious paralysis of the will which overwhelmed him as he sat in his palace and implored—in vain—his friend, the Government of the United States, to tell him what he should do.

Shortly after Khomeini's triumphant arrival in Tehran, a popular referendum approved the setting up of the Islamic Republic of Iran, and shortly after that another referendum approved the Constitution of the Republic—a very long and involved document drafted by an 'Assembly of [religious]

Experts' and approved by a constituent assembly. This course of events and the constitutional arrangements associated with it represents a curious amalgam of Khomeini's understanding of what a pure Muslim polity requires, and Western constitutional devices which, however much Khomeini and his Religious Experts must have disliked them, they still, for some reason, thought it necessary to include them in the Constitution. Indeed, the very idea of a polity governed by a man-made constitution is alien to Islam since, as Khomeini declared, all legislation comes from God. Again, if all true legislation is divinely sanctioned, what place is there for public opinion, popular plebiscites, and representative assemblies, and above all, for a republic—a republic, a *jumhuriyet*, being governed by the *jumhur*, the vulgar mass, a state of affairs deeply repugnant to Muslim religious authorities? Yet we find that the Constitution stipulates, under Principle 6, that:

The affairs of the country should be administered in the Islamic Republic of Iran by relying upon public opinion, expressed through elections, i.e. election of the president of the republic, deputies of the National Assembly, members of councils, and the like, or by plebiscite, anticipated for cases specified in other principles of this Constitution.

If, then, the Constitution realizes Khomeini's dream, the realization is not absolutely pure and unsullied. However, if one closely examines the document, one may conclude that the inclusion of the Western ideas is, so to speak, no more than the tribute which (Islamic) virtue somehow feels it has to pay to (Western) vice. For the rest, as is declared in the preamble, the Constitution was completed

on the eve of the fifteenth century of the advent of the great Prophet of God and the anniversary of the redemptive school of Islam in the hope that this century will be the century of a world rule by the heretofore oppressed and the complete defeat of the arrogant.

In the meantime, before this consummation, we read in Principle 5:

During the absence of the glorious Lord of the Age, may God grant him relief, he will be represented in the Islamic Republic of Iran as religious leader and Imam of the people by an honest, virtuous, well-informed, courageous, efficient administrator and religious jurist.

As is specified in detail further on in the Constitution, this religious jurist who represents the Hidden Imam is indeed the fulcrum on which the whole edifice is poised. Principle 107 declares that this leader 'will have charge of governing and all the responsibilities arising from it'. The leader assumes command of all the armed forces, appointing and dismissing the chief of the general staff and the commanders-in-chief of all the armed forces, and declaring war and mobilizing the troops. He approves the competence of candidates to the presidency of the Republic, confirms the presidency of the elected candidate, and dismisses him following an order of the Supreme Court charging him with violating his legal duties. The leader also appoints the members of the Supreme Court, and the jurists serving on the Council of Guardians (Principle 110). This Council of Guardians is composed of twelve members, who are responsible for 'guarding the precepts of Islam and the constitution'. Six of these are 'just and religious persons' appointed by the leader, and six are lawyers from different branches of the law, nominated by the High Council of the Judiciary and appointed by the National Consultative Assembly. It is the six 'just and religious persons' appointed by the leader who decide whether a law passed by the National Assembly contradicts 'Islamic decrees', and unless it is modified by the Assembly, it becomes null and void. Furthermore, the Council approves the credentials of the representatives elected to the Assembly (Principles 91, 93, and 96). The first leader, representing the Hidden Imam, was, of course, as Principle 107 puts it, 'the Great Ayatollah Imam Khomeini', with his 'high calling to the leadership of the revolution'. Following him, a leader or, failing one with the necessary qualifications, a Leadership Council, will be appointed by a council of experts and 'introduced to the people'. The membership and functioning of this council is regulated by a law prepared in the Council of Guardians and approved by the 'revolutionary leader' (Principles 107 and 108).

The absolute and ultimate authority of the religious jurist, as adumbrated in Khomeini's lectures of 1970, is thus spelled out and fully entrenched by device upon ingenious device which make it well nigh impossible for any other organ of the state, let alone the people at large, to control or challenge the appointment of the leader or any of his decisions.

This highly centralized and autocratic form of government is endowed by the Constitution with far-reaching powers over all aspects of social and economic life: the Government 'must create, for all individuals, the possibility of employment and equal opportunities for obtaining it' (Principle 28); the Government must provide housing, and also secure other 'basic needs', namely 'nourishment, clothing, hygiene, medical care, education and vocational training, and establishing a suitable environment for all to start a family' (Principles 30, 31, and 43). The Government also will provide 'systematic and sound planning' for the economy; it will control and administer 'all major industries, foreign trade, major mines, banking, insurance, power production, dams and major water-carrying networks, radio and television, postal, telegraph, and telephone system, air, sea, land, and railroad transportation, and,' Principle 44 adds for good measure, 'others similar to the above'. The private sector, the Principle generously decrees, consists of those portions of agriculture, animal husbandry, trade, and services which 'supplement' the governmental sector.

The highly centralized and all-embracing autocracy created by the Constitution of the Islamic Republic seemingly endows the leader and his subordinates with the most sweeping powers over the people of Iran. These powers, however, *ipso facto* impose crushing responsibilities which any government may find very difficult, if not impossible, to discharge. The leader, even though he repesents the Hidden Imam, is, after all, only a mortal man. His plans and dispositions will be exposed to the chances and changes, the vicissitudes, to which are subject all human hopes and aspirations. If the original leader and his successors purport to act in the name of the 'Glorious Lord of the Age', and their plans and decisions end in disappointment and failure, will this not throw into discredit the very religion which Khomeini wished to save from the ruinous clutch of the impious Pahlavi and that of the Satanic unbelievers, his masters? Fears such as these moved some of Khomeini's colleagues to feel doubts and misgivings over the course on which Iran was embarked after 1979. This was notably true of Ayatollah Kazem Shariatmadari, who had been one of Khomeini's defenders when he was being pursued by the Shah. Shariatmadari seems to have made his scepticism known. He was silenced. By an unheard-of procedure, he was stripped of his rank of Ayatollah, forced to confess to a conspiracy against the regime, and put until his death under house arrest. What the Islamic Republic turned out to be was a centralized absolutism in which Khomeini and his clerical followers tenaciously held on to the reins of power and, while the leader was still alive, intrigued and fought in his shadow for power and position. The old regime had been swiftly destroyed. Death sentences and imprisonments by the thousand meted out to its leading figures, and wholesale confiscations, made sure that none would dare to stand in the way of the new dispensation. Less than eighteen months after Khomeini's coming to power, Iran became engaged in a costly and destructive war, lasting eight years, with its neighbour Iraq. One of the reasons which must have tempted Iraq to make the attack which began the war was that Khomeini decimated, imprisoned, or cashiered the officers of the Iranian Army out of fear of their possible disloyalty. His actions thus contributed, perhaps decisively, to the catastrophe which has been so far the only notable event in the life of the Islamic Republic.

The establishment of the Islamic Republic of Iran, the military *coups d'état* in Egypt, Syria, and Iraq, and the regimes issuing from them, the destruction of the Lebanese Republic, and the mixed fortunes of constitutional and representative government in the Turkish Republic are the outcome, thus far, of one hundred and fifty years of tormented endeavour to discard the old ways, which have ceased to satisfy and to replace them with something modern, eye-catching, and attractive. The torment does not seem likely to end soon.

Notes

Chapter 1. The Legacy

1. See P. Crone and M. Hinds, *God's Caliph: Religious Authority in the First Centuries of Islam* (Cambridge, 1986).
2. Ali al-Qari, in *Sharh al-fiqh al-akbar*, quoted in I. Goldziher, *Introduction to Islamic Theology and Law* (Princeton, 1981), 182–3.
3. Quoted in S. G. Haim, 'Islam and the Theory of Arab Nationalism', in W. Z. Laqueur (ed.), *The Middle East in Transition* (London, 1958), 304–6.
4. Quoted in A. K. S. Lambton, *State and Government in Medieval Islam* (London, 1981), 124.
5. W. Madelung, 'Hisham ibn al-Hakam', in *Encyclopaedia of Islam*, ed. C. E. Bosworth, E. van Donzel, B. Lewis, and C. Pellat, 2nd edn. (Leiden, 1979–89); Lambton, *State and Government in Medieval Islam*, 261 n. 66.
6. K. Wittfogel, *Oriental Despotism* (New Haven, 1957).
7. A. H. Lybyer, *The Government of the Ottoman Empire in the Time of Suleiman the Magnificent* (Cambridge, Mass., 1913).
8. Lambton, *State and Government*, 143 n. 16.
9. Koran XI:29, here cited in B. Lewis's translation, in J. Schecht and C. E. Bosworth (eds.), *The Legacy of Islam*, 2nd edn. (Oxford, 1974), 180.

Chapter 2. The Modern World: Threat and Predicament

1. Quoted in E. Kedourie, *Arabic Political Memoirs* (London, 1974), 4.
2. Quoted in C. V. Findley, *Bureaucratic Reform in the Ottoman Empire* (Princeton, 1980), 116.
3. Khayr al-Din al-Tunisi, *The Surest Path: The Political Treatise of a Nineteenth-Century Statesman*, trans. L. C. Brown (Cambridge, Mass., 1967), 163.

4. Quoted in B. Lewis, *The Emergence of Modern Turkey* (London, 1961), 232.

Chapter 3. Constitutionalism and its Failure: I

1. F. Millingen, quoted in Ş. Mardin, *The Genesis of Young Ottoman Thought* (Princeton, 1962), 112.
2. Ibid. 378.
3. 'Persian Civilization', *The Contemporary Review*, London (1891), 238–44.
4. Koran III:103.
5. Cited in Abdul-Hadi Hairi, *Shiism and Constitutionalism in Iran* (Leiden, 1977), 181–2.
6. Cited in Suna Kili, *Kemalism* (Istanbul, 1969), 62–3.
7. Quoted in E. Kedourie, *Arabic Political Memoirs* (London, 1974), 4.
8. Lord Cromer, *Modern Egypt* (London, 1908), i. 84 ff.
9. Cited in R. Devereux, *The First Ottoman Constitutional Period* (Baltimore, 1963), 31 n. 21.
10. Ibid. 40.
11. Cited in Mardin, *The Genesis of Young Ottoman Thought*, 19.
12. Quoted in S. J. Shaw and E. K. Shaw, *History of the Ottoman Empire and Modern Turkey* (Cambridge, 1977), 213.
13. Quoted in V. R. Swenson, 'The Young Turk Revolution', Ph.D. thesis (Johns Hopkins University, 1960), 112–13.
14. Quoted in V. R. Swenson, 'The Military Rising in Istanbul 1909', *Journal of Contemporary History*, 5 (1970).
15. Quoted in Swenson, 'The Young Turk Revolution', 389.
16. Lord Curzon, *Persia and the Persian Question* (London, 1892), i. 398–9.
17. Quoted in Abdul-Hadi Hairi, *Shiism and Constitutionalism in Iran*, 26.
18. Lord Curzon, *Persia and the Persian Question*, i. 465.

Chapter 4. Turkey after 1919: The Failure of Constitutionalism?

1. Ahmed Shafiq Pasha, quoted in E. Kedourie, 'Egypt and the Caliphate, 1915–1952', *The Chatham House Version and Other Middle Eastern Studies* (London, 1970), 194.
2. Ibid. 207.

3. G. Lewis, 'Political Change in Modern Turkey since 1960', in *Aspects of Modern Turkey*, ed. W. Hale (London, 1976), 15.

4. Feroz Ahmad, *The Turkish Experience in Democracy* (London, 1977), 30.

5. Figures quoted in S. J. Shaw and E. K. Shaw, *History of the Ottoman Empire and Modern Turkey* (Cambridge, 1977), ii. 409.

6. 'Problems Unknown to Atatürk', *The Times*, 24 July 1962.

7. B. Lewis, *The Emergence of Modern Turkey*, (London, 1961), 458–9.

8. Quoted in D. Barchard, *Turkey and the West* (London, 1985), 20.

9. Quoted in Feroz Ahmad, *The Turkish Experience in Democracy*, 338.

10. Ibid. 339–40.

11. Ibid. 349.

12. C. H. Dodd, *The Crisis of Turkish Democracy*, 2nd. edn. (Huntingdon, 1990), 27.

13. *The Daily Telegraph*, 28 November 1989.

14. 'Transition to Democracy' in *State Democracy and the Military*, ed. M. Heper and A. Evin (London, 1988).

Chapter 5. Constitutionalism and its Failure: II

1. Kedourie, *The Chatham House Version and Other Middle Eastern Studies*, 86–8.

2. Lord Milner, *England in Egypt* (London, 1892).

3. Kedourie, *The Chatham House Version*, 118–19, 413.

4. 'Saʻd Zaghlul and the British' in Kedourie, *The Chatham House Version*, 123.

5. Ibid. 122–4.

6. Ibid. 132–3.

7. A. D. MacDonald, *Euphrates Exile* (London, 1936), 54–6.

8. Proceedings of the Permanent Mandates Commission, 20th Session.

9. Quoted in E. Kedourie, *The Chatham House Version*, 304.

10. Ibid. 222.

11. Ibid.

12. P. Seale, *The Struggle for Syria* (London, 1965), 32.

13. Khalid al-Azm, *Mudhakkirat (Memoirs)*, 2nd edn. (Beirut, 1973) i. 211–12.

14. Seale, *The Struggle for Syria*, 32.

15. Khalid al-Azm, *Memoirs*, ii. 271.

16. Cited in P. Rondot, *Les institutions politiques du Liban* (Paris, 1947), 80.

17. M. Chiha, 'Politique intérieure, Beirut 1964', quoted in *The Precarious Republic* ed. M. C. Hudson (New York, 1968), 92.

18. A. Soffer, 'Lebanon—Where Demography is the Core of Politics and Life', *Middle Eastern Studies* 22 (1986), 192–205.

19. Quoted in I. Rabinovich, *The War for Lebanon 1970–1983*, (Ithaca, 1984), 39.

20. Quoted in N. Kliot, 'The Collapse of the Lebanese State', *Middle Eastern Studies* 23 (1987), 54–74.

21. Quoted in E. Abrahamian, *Iran between Two Revolutions* (Princeton, 1982), 151.

22. Ibid. 138.

23. Ibid. 198.

24. Quoted in K. M. Dadkhah, 'The Oil Nationalization Movement, the British Boycott, and the Iranian Economy', in *Essays on the Economic History of the Middle East* ed. E. Kedourie and S. G. Haim (London, 1988), 128.

25. Abrahamian, *Iran between Two Revolutions*, 441.

Chapter 6. The Triumph of Ideological Politics

1. W. Wordsworth, 'French Revolution as it Appeared to Enthusiasts at its Commencement', 1804.

2. Quoted from Afghani's articles 'Le Mahdi' in *L'Intransigeant* (Paris, 1883), reproduced in E. Kedourie, *Afghani and 'Abduh: An Essay on Religious Unbelief and Political Activism in Modern Islam* (London, 1966), appendix ii.

3. Henri Rochefort, *Les aventures de ma vie* (Paris, 1897), iv. 345 ff.

4. E. Renan, *Discours et Conférences* (Paris, 1887), 402–3.

5. Quoted in Kedourie, *Afghani and Abduh*, 14.

6. *The New York Times*, 6 Sept. 1990.

7. T. Alp, *The Turkish and Pan-Turkish Ideal* (London, 1917), 42–3.

8. Cited in S. G. Haim, *Arab Nationalism* (Berkeley, 1962), 98–9.

9. Z. Gökalp, *Turkish Nationalism and Western Civilization*, trans. and ed. N. Berkes (London, 1959), 71–85.

10. Ibid.

11. B. Lewis, *The Emergence of Modern Turkey*, 348–97.

12. T. Alp, *Le Kemalisme* (Paris, 1937), chapter on 'The Restoration of Turkish History'.

13. Quoted in S. G. Haim, 'Islam and the Theory of Arab Nationalism', in *The Middle East in Transition*, ed. W. Z. Laqueur, 283.

14. Quoted in Haim, *Arab Nationalism*, 70–71.

15. Ibid. 71.

16. P. J. Vatikiotis, *Egypt from Muhammad Ali to Sadat*, 2nd edn. (London, 1980), 424–5.

17. T. Petran, *Syria* (London, 1972), 208.

18. M. Seurat, 'Les populations, l'état, et la société', in *La Syrie aujourd'hui*, ed. A. Raymond (Paris, 1980), 124–8.

19. P. Clawson, 'How Vulnerable is Iraq's Economy?', Research Memorandum No. 14 (Washington Institute for Near East Policy, 1990).

20. Quoted in E. Kedourie, *Arabic Political Memoirs*, 175.

21. Abd al-Karim Zuhur, Syrian Minister of National Economy, quoted in *BBC Summary of World Broadcasts* part vi, 27 June 1963, appendix e, 3–6.

22. Seale, *Asad of Syria*, 184.

23. Ibid. 150–1.

24. Quoted in Kedourie, *Arabic Political Memoirs*, 202.

25. Quoted in S. G. Haim, *Arab Nationalism*, 172–88.

26. Ibid. 167–71.

27. Quoted in Kedourie, *Arabic Political Memoirs*, 211.

28. Seale, *Asad of Syria*, 328–9.

29. Ibid. 334.

30. Mulla Ali Kani, quoted by Ali Reza Sheikholeslami, 'From Religious Accommodation to Religious Revolution' in *The State, Religion, and Ethnic Politics: Afghanistan, Iran, and Pakistan*, ed. A. Banuazizi and M. Weiner (Syracuse, 1986), 239.

31. 'Islamic Government', in *Islam and Revolution: Writings and Declarations of Imam Khomeini*, trans. Hamid Algar (Berkeley, 1981), 68.

32. Ibid. 60.

Further Reading

A GOOD introduction to the civilization of Islam from its beginnings until modern times is *The World of Islam*, edited by Bernard Lewis (London, 1976); readers of this book will find particularly useful the chapters by Lewis on 'The Faith and The Faithful', by Oleg Grabar on 'Cities and Citizens', by Norman Itzkowitz on 'The Ottoman Empire', and by Elie Kedourie on 'Islam Today'. *The Legacy of Islam*, edited by Joseph Schacht and C. E. Bosworth, 2nd edn. (Oxford, 1974) includes two substantial chapters: 'Politics and War' by Bernard Lewis, and 'Law and the State' by Joseph Schacht and A. K. S. Lambton, which are of particular relevance in elucidating the traditional Islamic political and legal context in which modernization must be understood.

Two classic works describe the characteristics of the modernizing Middle Eastern State in the nineteenth century: Sir Charles Eliot, *Turkey in Europe*, 2nd edn. (London 1907), and Lord Curzon's *Persia and the Persian Question*, 2 vols. (London, 1892).

Two general studies of constitutionalism and party politics in the Middle East are included in Elie Kedourie, *Arabic Political Memoirs and Other Studies* (London, 1974): 'The Fate of Constitutionalism in the Middle East' and 'Political Parties in the Arab World'.

The origin, course, and vicissitudes of constitutionalism in the Ottoman Empire are elucidated in H. W. V. Temperley, 'British Policy Towards Parliamentary Rule and Constitutionalism in Turkey, 1830–1914', in *Cambridge Historical Journal* (1933); in Şerif Mardin, *The Genesis of Young Ottoman Thought* (Princeton, 1962); in E. E. Ramsaur Jr., *The Young Turks: Prelude to the Revolution of 1908* (Princeton, 1957); and most notably in Bernard Lewis, *The Emergence of Modern Turkey* (London, 1961 and subsequent editions) which covers concisely and authoritatively political and intellectual change during the Ottoman *tanzimat*, the Hamidian and Young Turk periods, as well as the Kemalist and post-Kemalist republic. Stanford J. Shaw and Ezel Kural Shaw, *History of the Ottoman Empire and Modern Turkey*, II: *Reform, Revolution and Republic: The Rise of Modern Turkey 1808–1975* (Cambridge, 1977), a detailed history, carries the story of Turkish politics down to its terminal date, 1975.

Edward G. Browne, *The Persian Revolution of 1905–1909* (Cambridge, 1910) is a detailed account of the events of 1906–7 in Persia, their antecedents and sequels. The very fervour of its enthusiasm for the Constitutional Revolution enhances its readability, making it, as well, a striking and eloquent specimen of the literature of political commitment. Peter Avery, *Modern Iran* (London, 1965) provides a concise account of the political history of Iran, particularly after the First World War and during the reign of Reza Shah Pahlavi and the first two decades or so of that of his son Mohammed Reza Shah. Ervand Abrahamian, *Iran Between Two Revolutions* (Princeton, 1982), analyses political and social forces in Iran before, and after, 1906 and provides a valuable account of Iranian politics after the abdication of Reza Shah and the Allied occupation in 1941. The work discusses the fluid character of Iranian politics in the period when Reza Shah's successor had not yet succeeded in imposing his authority, and the manner in which Mohammed Reza did at last succeed in establishing and consolidating the autocracy which he continued to exercise until his downfall in 1979.

P. J. Vatikiotis, *The Modern History of Egypt*, 4th edn. (London, 1991) surveys authoritatively the transformation of an Ottoman province into an autonomous entity, and at length into a sovereign state which changed from a monarchy into a republic. The same author's *Nasser and his Generation* (London, 1978) gives an account of the antecedents and record of the man who was chiefly responsible for ending the rule of Muhammad Ali's dynasty in 1952, and substituting for it his own eighteen-year long autocracy. Elie Kedourie, *The Chatham House Version and Other Middle Eastern Studies* (London, 1970, paperback edn, London 1984), includes three studies dealing with issues of Egyptian politics under the

monarchy which are fundamental to understanding the vicissitudes of Egyptian constitutionalism: 'Sa'd Zaghlul and the
British', 'The Genesis of the Egyptian Constitution of 1923'
and 'Egypt and the Caliphate, 1915–1952'.

The Chatham House Version, includes two studies relating to
Iraq under the monarchy: 'The Kingdom of Iraq: A Retrospect'
and 'Minorities'. This latter study also discusses the clash
between the Armenians and the Ottoman Empire at the end of
the nineteenth century—a clash illustrating the consequences
of the spread of ideological politics in the Middle East, which
is discussed in Chapter 6 above. The titles of two books,
Marion Farouk-Sluglett and Peter Sluglett, *Iraq Since 1958:
From Revolution to Dictatorship* (London, 1987) and Samir al-
Khalil, *Republic of Fear: The Politics of Modern Iraq* (Berkeley, 1989) are self-explanatory: they delineate the fortunes and
the characteristics of the Iraqi polity in the three decades which
followed the destruction of the monarchy.

Khalil is particularly concerned with the Ba'thist regime
which came to power in Iraq in 1968. A companion volume, in
a manner of speaking, is Patrick Seale's *Asad of Syria* (London,
1988). Just as Samir al-Khalil describes the Iraqi regime set up
and controlled by Saddam Husayn al-Tikriti, so Patrick Seale,
in his very different way, delineates the (quite similar) features
of the Syrian regime set up and controlled by Saddam Husayn's
fellow Ba'thist, Hafiz al-Asad. The same author's *Struggle for
Syria* (London, 1965) and Tabitha Petran's *Syria* (London,
1972), deal with Syrian politics after the end of the French
Mandate and S. H. Longrigg, *Syria and the Lebanon under the
French Mandate* (London, 1958) details the politics of the
period between 1920 and 1946.

As is clear from its title, Longrigg's book takes in Lebanese
politics under the French mandate. A work which puts this
period into the context of Lebanese history is Kamal Salibi,
The Modern History of Lebanon (London, 1965). Itamar
Rabinovich, *The War for Lebanon 1970–1983* (Ithaca and
London, 1984) discusses (in its first four chapters) the events
which ended by destroying the Lebanese Republic as established by the French Mandatory.

The introduction to Elie Kedourie, *Nationalism in Asia and Africa* (New York, 1970) deals with the spread of the ideological style of politics outside Europe *inter alia* in the Ottoman Empire and the Arab world; the work includes essays by Turkish and Arab ideologists of nationalism. The origins of ideological politics in the Middle East are investigated in Elie Kedourie, *Afghani and 'Abduh: An Essay on Religious Unbelief and Political Activism in Modern Islam* (London, 1966).

Sylvia G. Haim, *Arab Nationalism* (Berkeley, 1962) discusses, in a long introduction, writings by Arab nationalists collected and translated by her. The essay 'Arabic Political Memoirs' included in Elie Kedourie, *Arabic Political Memoirs and Other Studies*, examines the progress and effects of ideological politics in the Arab world after the Second World War. Islamic 'fundamentalism', the ideology which competes with nationalism, is discussed, in its Sunni variety, by Richard P. Mitchell, *The Society of the Muslim Brothers* (London, 1969). Its Shiite variety is expounded in the writings of Ayatollah Ruhallah Khomeini, collected and translated by Hamid Algar, *Islam and Revolution: Writings and Declarations of Imam Khomeini* (Berkeley, 1981). The character of his doctrinal innovation is discussed in Elie Kedourie, *Islam in the Modern World* (New York, 1980), chapter 3: 'Rule and Religion in Iran'.

Index

OXFORD

MORE OXFORD PAPERBACKS

Details of a selection of other Oxford Paperbacks follow. A complete list of Oxford Paperbacks, including The World's Classics, Twentieth-Century Classics, OPUS, Past Masters, Oxford Authors, Oxford Shakespeare, and Oxford Paperback Reference, is available in the UK from the General Publicity Department, Oxford University Press (RS), Walton Street, Oxford, OX2 6DP.

In the USA, complete lists are available from the Paperbacks Marketing Manager, Oxford University Press, 200 Madison Avenue, New York, NY 10016.

Oxford Paperbacks are available from all good bookshops. In case of difficulty, customers in the UK can order direct from Oxford University Press Bookshop, 116 High Street, Oxford, Freepost, OX1 4BR, enclosing full payment. Please add 10 per cent of the published price for postage and packing.

POLITICS IN OXFORD PAPERBACKS

Oxford Paperbacks offers incisive and provocative studies of the political ideologies and institutions that have shaped the modern world since 1945.

GOD SAVE ULSTER!

The Religion and Politics of Paisleyism

Steve Bruce

Ian Paisley is the only modern Western leader to have founded his own Church and political party, and his enduring popularity and success mirror the complicated issues which continue to plague Northern Ireland. This book is the first serious analysis of his religious and political careers and a unique insight into Unionist politics and religion in Northern Ireland today.

Since it was founded in 1951, the Free Presbyterian Church of Ulster has grown steadily; it now comprises some 14,000 members in fifty congregations in Ulster and ten branches overseas. The Democratic Unionist Party, formed in 1971, now speaks for about half of the Unionist voters in Northern Ireland, and the personal standing of the man who leads both these movements was confirmed in 1979 when Ian R. K. Paisley received more votes than any other member of the European Parliament. While not neglecting Paisley's 'charismatic' qualities, Steve Bruce argues that the key to his success has been his ability to embody and represent traditional evangelical Protestantism and traditional Ulster Unionism.

'original and profound . . . I cannot praise this book too highly.'
Bernard Crick, *New Society*

Also in Oxford Paperbacks:

Freedom Under Thatcher   Keith Ewing and Conor Gearty
Strong Leadership   Graham Little
The Thatcher Effect   Dennis Kavanagh and Anthony Seldon

HISTORY IN OXFORD PAPERBACKS

Oxford Paperbacks' superb history list offers books on a wide range of topics from ancient to modern times, whether general period studies or assessments of particular events, movements, or personalities.

THE STRUGGLE FOR
THE MASTERY OF EUROPE 1848-1918

A. J. P. Taylor

The fall of Metternich in the revolutions of 1848 heralded an era of unprecedented nationalism in Europe, culminating in the collapse of the Hapsburg, Romanov, and Hohenzollern dynasties at the end of the First World War. In the intervening seventy years the boundaries of Europe changed dramatically from those established at Vienna in 1815. Cavour championed the cause of *Risorgimento* in Italy; Bismarck's three wars brought about the unification of Germany; Serbia and Bulgaria gained their independence courtesy of the decline of Turkey—'the sick man of Europe'; while the great powers scrambled for places in the sun in Africa. However, with America's entry into the war and President Wilson's adherence to idealistic internationalist principles, Europe ceased to be the centre of the world, although its problems, still primarily revolving around nationalist aspirations, were to smash the Treaty of Versailles and plunge the world into war once more.

A. J. P. Taylor has drawn the material for his account of this turbulent period from the many volumes of diplomatic documents which have been published in the five major European languages. By using vivid language and forceful characterization, he has produced a book that is as much a work of literature as a contribution to scientific history.

'One of the glories of twentieth-century writing.' *Observer*

Also in Oxford Paperbacks:

Portrait of an Age: Victorian England G. M. Young
Germany 1866–1945 Gorden A. Craig
The Russian Revolution 1917–1932 Sheila Fitzpatrick
France 1848–1945 Theodore Zeldin

WOMEN'S STUDIES FROM
OXFORD PAPERBACKS

Ranging from the *A–Z of Women's Health* to *Wayward Women: A Guide to Women Travellers*, Oxford Paperbacks cover a wide variety of social, medical, historical, and literary topics of particular interest to women.

DESTINED TO BE WIVES
The Sisters of Beatrice Webb
Barbara Caine

Drawing on their letters and diaries, Barbara Caine's fascinating account of the lives of Beatrice Webb and her sisters, the Potters, presents a vivid picture of the extraordinary conflicts and tragedies taking place behind the respectable façade which has traditionally characterized Victorian and Edwardian family life.

The tensions and pressures of family life, particularly for women; the suicide of one sister; the death of another, probably as a result of taking cocaine after a family breakdown; the shock felt by the older sisters at the promiscuity of their younger sister after the death of her husband are all vividly recounted. In all the crises they faced, the sisters formed the main network of support for each other, recognizing that the 'sisterhood' provided the only security in a society which made women subordinate to men, socially, legally, and economically.

Other women's studies titles:

A–Z of Women's Health   Derek Llewellyn-Jones
'Victorian Sex Goddess': Lady Colin Campbell and the Sensational Divorce Case of 1886   G. H. Fleming
Wayward Women: A Guide to Women Travellers
Jane Robinson
Catherine the Great: Life and Legend   John T. Alexander

LAW FROM OXFORD PAPERBACKS

Oxford Paperbacks's law list ranges from introductions to the English legal system to reference books and in-depth studies of contemporary legal issues.

INTRODUCTION TO ENGLISH LAW
Tenth Edition

William Geldart
Edited by D. C. M. Yardley

'Geldart' has over the years established itself as a standard account of English law, expounding the body of modern law as set in its historical context. Regularly updated since its first publication, it remains indispensable to student and layman alike as a concise, reliable guide.

Since publication of the ninth edition in 1984 there have been important court decisions and a great deal of relevant new legislation. D. C. M. Yardley, Chairman of the Commission for Local Administration in England, has taken account of all these developments and the result has been a considerable rewriting of several parts of the book. These include the sections dealing with the contractual liability of minors, the abolition of the concept of illegitimacy, the liability of a trade union in tort for inducing a person to break his/her contract of employment, the new public order offences, and the intent necessary for a conviction of murder.

Other law titles:

Freedom Under Thatcher: Civil Liberties in Modern Britain
Keith Ewing and Conor Gearty
Doing the Business   Dick Hobbs
Judges   David Pannick
Law and Modern Society   P. S. Atiyah

RELIGION AND THEOLOGY FROM OXFORD PAPERBACKS

The Oxford Paperbacks's religion and theology list offers the most balanced and authoritative coverage of the history, institutions, and leading figures of the Christian churches, as well as providing in-depth studies of the world's most important religions.

MICHAEL RAMSEY
A Life

Owen Chadwick

Lord Ramsey of Canterbury, Archbishop of Canterbury from 1961 to 1974, and one of the best-loved and most influential churchmen of this century, died on 23 April 1988.

Drawing on Dr Ramsey's private papers and free access to the Lambeth Palace archive, Owen Chadwick's biography is a masterly account of Ramsey's life and works. He became Archbishop of Canterbury as Britain entered an unsettled age. At home he campaigned politically against racialism and determined to secure justice and equality for immigrants. In Parliament he helped to abolish capital punishment and to relax the laws relating to homosexuality. Abroad he was a stern opponent of apartheid, both in South Africa and Rhodesia. In Christendom at large he promoted a new spirit of brotherhood among the churches, and benefited from the ecumenism of Popes John XXIII and Paul VI, and the leaders of the Orthodox Churches of Eastern Europe.

Dr Ramsey emerges from this book as a person of much prayer and rock-like conviction, who in an age of shaken belief and pessimism was an anchor of faith and hope.

Other religion and theology titles:

John Henry Newman: A Biography   Ian Ker
John Calvin   William Bouwsma
A History of Heresy   David Christie-Murray
The Wisdom of the Saints   Jill Haak Adels

SCIENCE IN OXFORD PAPERBACKS

Oxford Paperbacks offers a challenging and controversial list of science and mathematics books, ranging from theories of evolution to analyses of the latest micro-technology, from studies of the nervous system to advice on teenage health.

THE AGES OF GAIA

A Biography of Our Living Earth

James Lovelock

In his first book, *Gaia: A New Look at Life on Earth* (OPB, 1982), James Lovelock proposed a startling new theory of life. Previously it was accepted that plants and animals evolve on, but are distinct from, an inanimate planet. Gaia maintained that the Earth, its rocks, oceans, and atmosphere, and all living things are part of one great organism, evolving over the vast span of geological time. Much scientific work has since confirmed Lovelock's ideas.

In this new book, Lovelock elaborates the basis of a new and unified view of the earth and life sciences, discussing recent scientific developments in detail: the greenhouse effect, acid rain, the depletion of the ozone layer and the effects of ultraviolet radiation, the emission of CFCs, and nuclear power. He demonstrates the geophysical interaction of atmosphere, oceans, climate, and the Earth's crust, regulated comfortably for life by living organisms using the energy of the sun.

'Open the cover and bathe in great draughts of air that excitingly argue the case that "the earth is alive".' David Bellamy, *Observer*

'Lovelock deserves to be described as a genius.' *New Scientist*

'He is to science what Gandhi was to politics.' Fred Pearce, *New Scientist*

Also in Oxford Paperbacks:

What is Ecology?   Denis Owen
The Selfish Gene 2/e   Richard Dawkins
The Sacred Beetle and Other Great Essays in Science
Chosen and introduced by Martin Gardner

MEDICINE IN OXFORD PAPERBACKS

Oxford Paperbacks offers an increasing list of medical studies and reference books of interest to the specialist and general reader alike, including The Facts series, authoritative and practical guides to a wide range of common diseases and conditions.

CONCISE MEDICAL DICTIONARY
Third Edition

Written without the use of unnecessary technical jargon, this illustrated medical dictionary will be welcomed as a home reference, as well as an indispensible aid for all those working in the medical profession.

Nearly 10,000 important terms and concepts are explained, including all the major medical and surgical specialities, such as gynaecology and obstetrics, paediatrics, dermatology, neurology, cardiology, and tropical medicine. This third edition contains much new material on pre-natal diagnosis, infertility treatment, nuclear medicine, community health, and immunology. Terms relating to advances in molecular biology and genetic engineering have been added, and recently developed drugs in clinical use are included. A feature of the dictionary is its unusually full coverage of the fields of community health, psychology, and psychiatry.

Each entry contains a straightforward definition, followed by a more detailed description, while an extensive crossreference system provides the reader with a comprehensive view of a particular subject.

Also in Oxford Paperbacks:

Drugs and Medicine   Roderick Cawson and Roy Spector
Travellers' Health: How to Stay Healthy Abroad 2/e
Richard Dawood
I'm a Health Freak Too!
Aidan Macfarlane and Ann McPherson
Problem Drinking   Nick Heather and Ian Robertson